KU-651-183

A FUNCTIONAL BIOLOGY OF MARINE GASTROPODS

FUNCTIONAL BIOLOGY SERIES

General Editor: Peter Calow, Department of Zoology, University of Sheffield

A Functional Biology of Free-living Protozoa
Johanna Laybourn-Parry

A Functional Biology of Sticklebacks
R.J. Wootton

A Functional Biology of Marine Gastropods
Roger N. Hughes

A Functional Biology of Nematodes
David A. Wharton

A Functional Biology of Marine Gastropods

Roger N. Hughes

School of Animal Biology,
University College of North Wales, Bangor.

CROOM HELM
London & Sydney

Croom Helm Ltd, Provident House, Burrell Row,
Beckenham, Kent BR3 1AT
Croom Helm Australia Pty Ltd, Suite 4, 6th Floor,
64-76 Kippax Street, Surry Hills, NSW 2010, Australia

British Library Cataloguing in Publication Data

Hughes, Roger N.
A functional biology of marine gastropods.
1. Gasteropoda
I. Title
594'.3 QL430.4

ISBN 0-7099-3746-6

Typeset in Times Roman by
Leaper & Gard Ltd, Bristol, England
Printed and bound in Great Britain by
Biddles Ltd, Guildford and King's Lynn

CONTENTS

To Helen, Ruth and Anne

PREFACE

All scientific evidence suggests that, whatever the precise nature of their origins, living organisms have evolved as devices that protect and replicate genes (Dawkins, 1976). To accomplish this, an organism must acquire materials and energy from its environment, using them to promote its own survival, growth and reproduction. In doing so, the organism influences others, perhaps affecting their survival, growth and reproduction. This book examines the functional aspects of marine gastropods that enable them to succeed as processors of materials and energy, and considers some of their effects on other organisms.

Anatomy and taxonomy are not dealt with comprehensively, but technical terms are defined and all genera mentioned in the text are listed under their taxonomic headings in the appendix. Chapter 1 gives an elementary classification and describes basic features of gastropods that may help the reader to appreciate fully the rest of the text.

Fitness, in an evolutionary sense, is gained by increasing the relative abundance of parental genes among the populations of subsequent generations. Depending on ecological circumstances, this may be achieved by rapid body growth and prodigious reproduction, by slower growth, more modest proliferation and greater longevity, or by various degrees and combinations of these traits. Whatever the quantitative details of genetic propagation, it is accomplished by somatic acquisition of materials and energy from the environment. Except for a few opisthobranchs that derive photosynthetic energy from symbiotic chloroplasts, both materials and energy are assimilated by gastropods from their food. Supplies of materials and energy are strongly correlated for carnivores, which may be able to meet all their nutritional requirements from a single type of prey. The correlation is less good among plant foods, some of which may be energetically rich but lacking in particular nutrients, making it necessary for the herbivore to broaden its diet.

Energy, nevertheless, is a dominant resource obtained from all foods and is necessary to drive all biological processes. There is therefore justification for conceptually adopting the energy maximisation premise that organisms function and behave in ways that maximise the ratio of energetic gain to cost (Townsend and Calow,

1981). Maximising this ratio automatically maximises the energy available for growth and reproduction. Genetic fitness, however, also depends on survival, which sometimes necessitates reducing the energetic gain relative to the cost. The biological activities of an organism therefore may be expected to reflect balanced investments that optimise productivity and survivorship.

Appreciation of these investments may be facilitated by compartmentalising biological processes into an energy budget. Following the standard format of the energy budget (Calow, 1981), ingested energy is balanced by absorbed and defecated energy. Absorbed energy is used in metabolism, excretion, secretion, growth and reproduction. By considering compartments of the energy budget, Chapters 2 to 6 attempt to present an evolutionarily rational illustration of the functional biology of marine gastropods. Their interactions with other organisms are considered at the community level and on a geographical scale in Chapter 7.

Roger N. Hughes

ACKNOWLEDGEMENTS

I thank Professor Peter Calow, Drs Brian L. Bayne, Derek A. Dorsett, David W. Phillips and Norman W. Runham for kindly reading the first draft of the manuscript and for suggesting many ways of improving it. I did not always take their advice and any deficiencies remain my own responsibility. I am grateful to the many authors who allowed me to use their published figures as a basis for the illustrations. Special thanks are extended to my wife, Helen, for typing the manuscript and for sharing my interest in marine gastropods.

FUNCTIONAL BIOLOGY SERIES: FOREWORD

General Editor: Peter Calow, Department of Zoology, University of Sheffield, England

The main aim of this series will be to illustrate and to explain the way organisms 'make a living' in nature. At the heart of this — their *functional biology*— is the way organisms acquire and then make use of resources in metabolism, movement, growth, reproduction, and so on. These processes will form the fundamental framework of all the books in the series. Each book will concentrate on a particular taxon (species, family, class or even phylum) and will bring together information on the form, physiology, ecology and evolutionary biology of the group. The aim will be not only to describe *how* organisms work, but also to consider *why* they have come to work in that way. By concentrating on taxa which are well known, it is hoped that the series will not only illustrate the success of selection, but also show the constraint imposed upon it by the physiological, morphological and developmental limitations of the groups.

Another important feature of the series will be its *organismic orientation.* Each book will emphasis the importance of functional *integration* in the day-to-day lives and the evolution of organisms. This is crucial since, though it may be true that organisms can be considered as collections of gene-determined traits, they nevertheless interact with their environment as integrated wholes and it is in this context that individual traits have been subjected to natural selection and have evolved.

The key features of the series are, therefore:

(1) Its emphasis on whole organisms as integrated, resource-using systems.
(2) Its interest in the way selection and constraints have moulded the evolution of adaptations in particular taxonomic groups.
(3) Its bringing together of physiological, morphological, ecological and evolutionary information.

P. Calow

1 DESIGN

1.1 Introduction

About 22000 species of gastropod are known from marine habitats and they dominate the largest, most diverse class in the phylum Mollusca (Figure 1.1A). Gastropods range from snails with robust shells and slugs without external shells, to vermiform endoparasites (Table 1.1). They are classified into three main groups: the prosobranchs, opisthobranchs and pulmonates, of which the first is the largest (Figure 1.1A).

Prosobranchs are placed in three orders, commonly called Archaeogastropoda, Mesogastropoda and Neogastropoda, but other names are sometimes used. Archaeogastropods (Plate 1.1) are evolutionarily the most primitive, and neogastropods the most advanced, although some predatory mesogastropods are also highly advanced. Coiling of the gastropod visceral mass has tended to suppress sets of organs on the right-hand side (section 1.7), but archaeogastropods retain varying degrees of bilateral symmetry and all of them possess a heart with two auricles (diotocardian) and a gill, or pair of gills, with filaments on either side of its central axis (bipectinate). Many archaeogastropods retain the ancestral fringe of tentacles on a flap of skin (mantle) encircling the upper foot.

Meso- and neogastropods (Plate 1.1) have lost certain organs on the right-hand side. The heart has only one auricle (monotocardian) and the gill has filaments only on one side of its central axis (monopectinate). Tentacles are usually reduced to the pair on the head (cephalic tentacles), although others may have arisen secondarily in a few species. Ganglia are more closely grouped around the oesophagus (nerve ring) than they are in the archaeogastropods (section 1.9). Mesogastropods constitute a large (Figure 1.1B), diverse assemblage ranging from herbivores and detritivores bearing more resemblance to archaeogastropods, to predators bearing a closer resemblance to neogastropods.

Neogastropods are distinguished from mesogastropods by the radula, which is specialised for predation and almost never has more than three teeth per row, and by several anatomical features

Figure 1.1.: A. Estimated Numbers of Living Species in the Major Molluscan Taxa. The Gastropoda (black bars) account for 80% of all molluscan species. B. Estimated Numbers of Living Species in the Three Main Taxonomic Groups within the Prosobranchia. Archaeogastropods account for 17%, mesogastropods 54% and neogastropods 29% of all prosobranchs. Estimates from different sources vary; the estimated total number of prosobranchs in A (20 000) is probably more accurate than that in B (12 808). Relative abundances of different taxa displayed in the histograms, however, are more reliable. From Boss (1971)

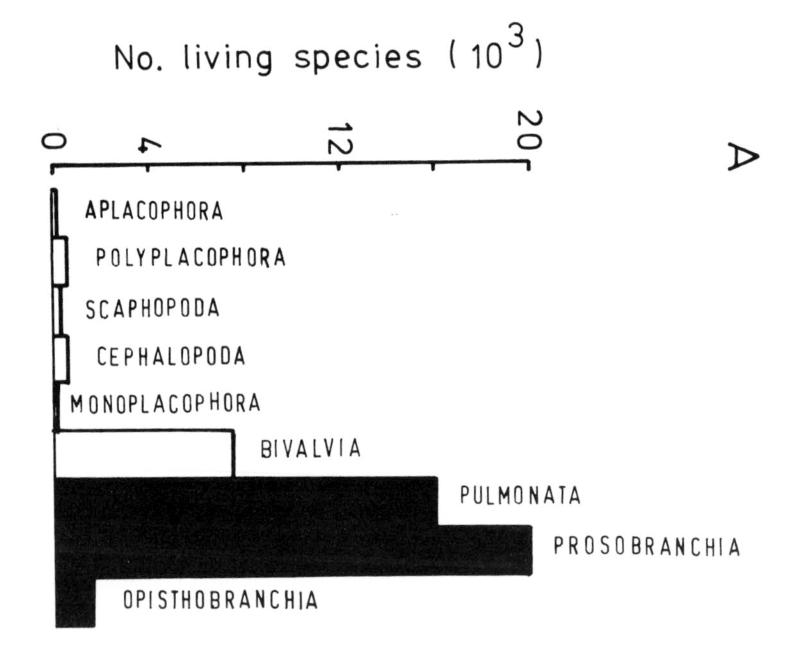

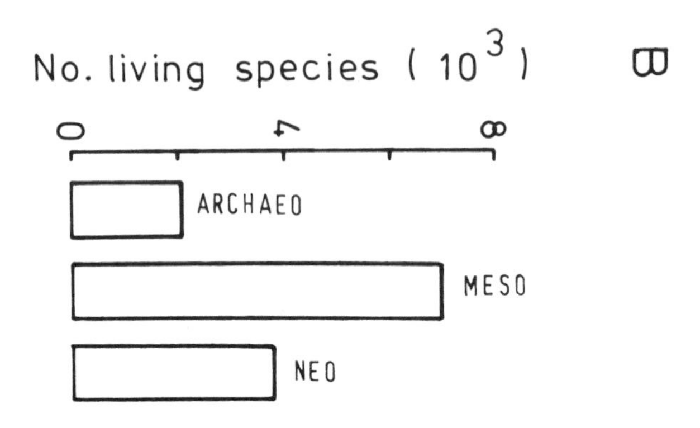

Table 1.1: Elementary Classification of Gastropods

The *phylum* **Mollusca** contains the *classes* Monoplacophora (rare, primitive deep-sea limpet-like molluscs); Amphineura (chitons); Gastropoda (snails, slugs, limpets); Scaphopoda (marine benthic molluscs with delicate tusk-shaped shells); Bivalvia (clams, scallops, mussels); and Cephalopoda (octopus, cuttlefish, squid).

The *class* **Gastropoda** is comprised of three *subclasses:* Prosobranchia (marine and a few freshwater snails); Opisthobranchia (marine slugs and some snails); and Pulmonata (terrestrial and most freshwater snails, terrestrial slugs, and some marine limpets). These subclasses contain the following orders.

Subclass **Prosobranchia**

Order **Archaeogastropoda**: most marine limpets, top shells and nerites; usually grazers of algal films.

Order **Mesogastropoda**: a huge, diverse group of snails including grazers, e.g. littorines (winkles), strombids, cowries; deposit-feeders, e.g. hydrobiids (mud-snails); filter-feeders. e.g. calyptraeaceans (slipper-limpets), vermetids (worm-shells), turritellids; predators, e.g. naticids (moon-shells), cassids (helmet-shells), cymatiids (tritons); and parasites, e.g. eulimids (ectoparasites), entoconchids (endoparasites).

Order **Neogastropoda**: a group of highly specialised predators, e.g. muricids (dogwhelks, oyster-drills, drupes, murex-shells), buccinids (whelks), nassariids (nassa mud-snails), olivids (olive-shells) and conids (cone-shells).

Subclass **Opisthobranchia**

Order **Bullomorpha**: slug-like forms with an external bubble-shaped shell; some are grazers, others are carnivores.

Order **Pyramidellomorpha**: small ectoparasitic snails.

Order **Thecosomata**: planktonic filter-feeding snails.

Order **Gymnosomata**: planktonic predatory snails.

Order **Aplysiomorpha**: sea-hares; algal grazers.

Order **Pleurobranchomorpha**: sea-slugs; grazers of benthic colonial animals.

Order **Acochlidiacea**: minute sea-slugs living among sand grains.

Order **Sacoglossa**: sea-slugs; specialised algal herbivores.

Order **Nudibranchia**: sea-slugs typically with numerous dorsal finger-like projections (cerata); specialised grazers on sedentary invertebrates.

Subclass **Pulmonata**

Order **Systellommatophora**: intertidal slugs; grazers of algal films.

Order **Basommatophora**: certain marine limpets, numerous freshwater snails and limpets.

Order **Stylommatophora**: terrestrial slugs and snails.

of the alimentary canal: the salivary gland ducts do not pass through the nerve ring as they do in mesogastropods; the glandular tissue lining the oesophagus (oesophageal gland) behind the nerve ring in mesogastropods is, in most neogastropods, collected into a separate gland (Leiblein's gland) that empties into the oesophagus via a short duct, and the stomach is reduced to a simple sac lacking complex ciliary sorting areas. The ganglia are even more centralised about the nerve ring than in most mesogastropods.

Plate 1.1: Diversity of Form in Marine Gastropods. Size denotes the main axis in each case. Collection localities are given. Photographs by E.W. Pritchard and D.J. Roberts

ARCHAEOGASTROPODS

A. *Haliotis midae*, 140 mm (Cape, South Africa)

B. *Patella barbara*, 60 mm (Cape, South Africa)

C. *Nerita peloronta*, 38 mm (West Indies)

D. *Trochus dentatus*, 80 mm (Red Sea)

E. *Turbo sarmaticus*, 75 mm (Cape, South Africa)

MESOGASTROPODS

F. *Cerithium echinatum*, 45 mm (Indian Ocean)

G. *Serpulorbis imbricatus*, 40 mm (Hong Kong)

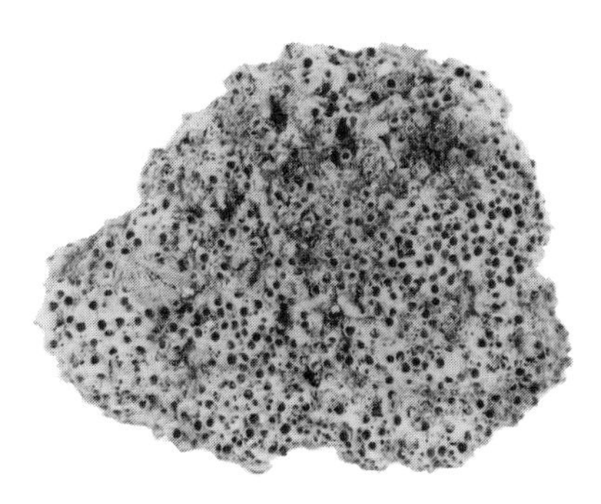

H. *Dendropoma corallinaceum*, aggregation 100 mm (Cape, South Africa)

I. *Strombus costatus*, 130 mm (West Indies)

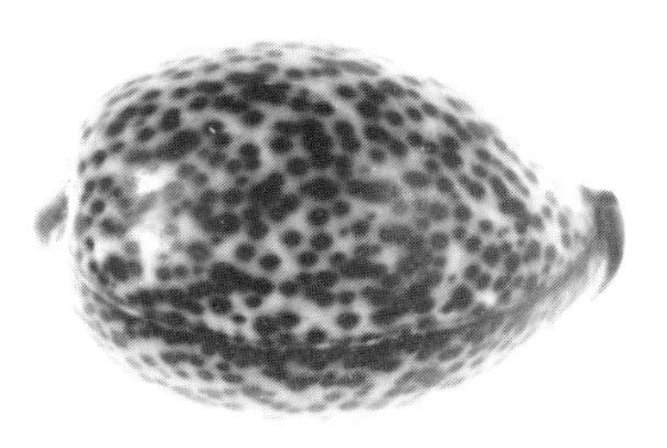

J. *Cypraea pantherina*, 72 mm (Red Sea)

MESOGASTROPODS continued

K. *Bursa granularis*, 50 mm (Red Sea)

L. *Cassis tuberosa*, 180 mm (West Indies)

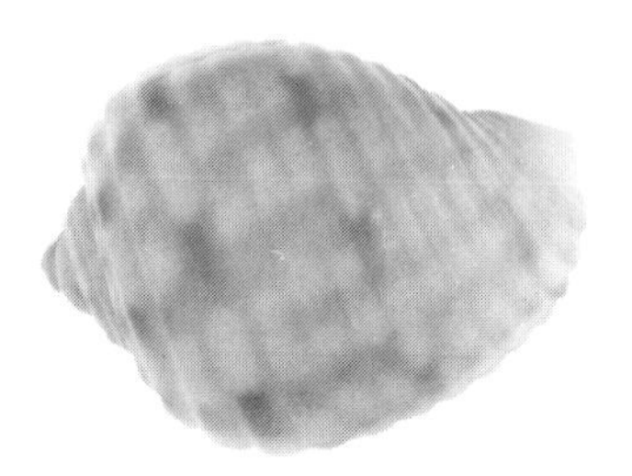

M. *Tonna perdix*, 31 mm (Indian Ocean)

N. *Natica unifasciata*, 32 mm (Pacific coast, Panama)

NEOGASTROPODS

O. *Nassarius albescens*, 27 mm (Indian Ocean)

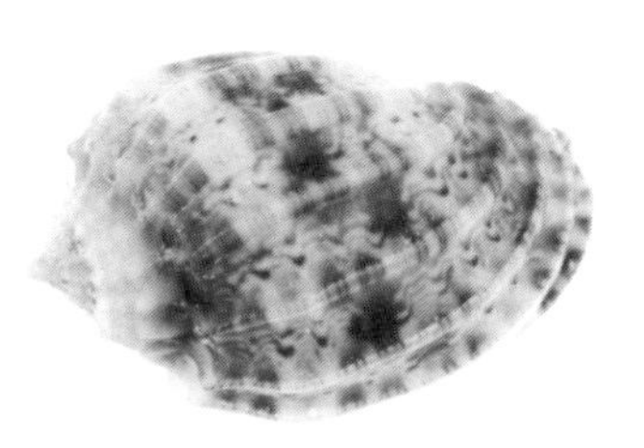

P. *Harpa amouretta*, 30 mm (Indian Ocean)

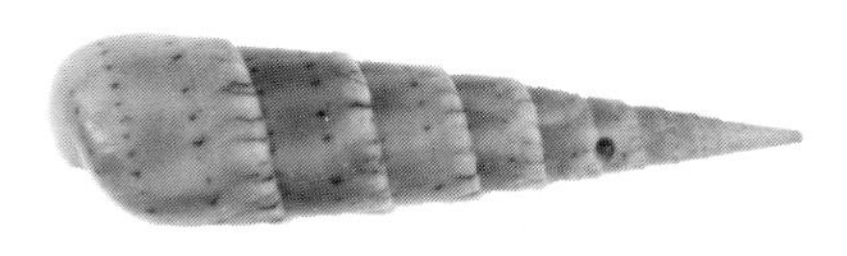

Q. *Terebra crenulata*, 90 mm (Red Sea)

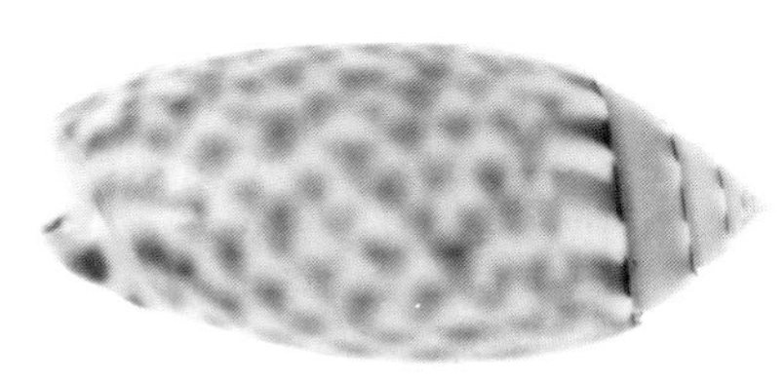

R. *Oliva annulata*, 52 mm (Indian Ocean)

S. *Mitra mitra*, 61 mm (Indian Ocean)

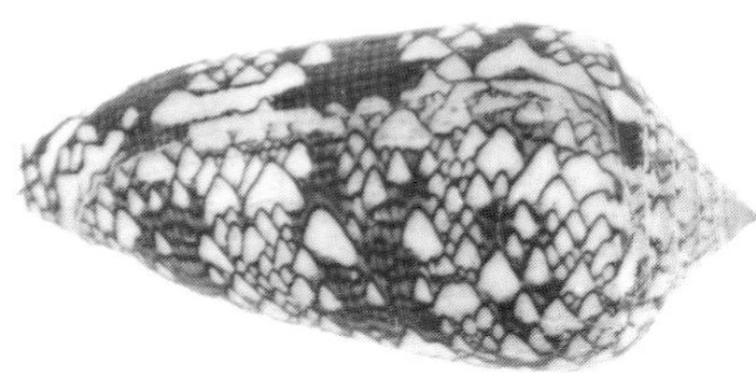

T. *Conus textile*, 80 mm (Red Sea)

OPISTHOBRANCHS

U. *Casella atromarginata*, 40 mm (Indian Ocean)

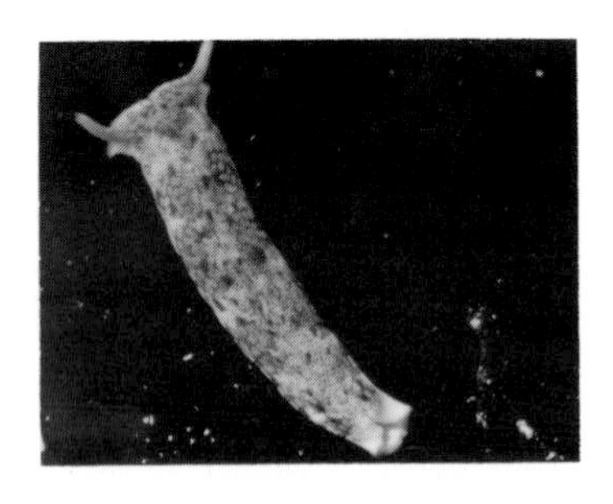

V. *Placobranchus ocellatus*, 30 mm (Indian Ocean)

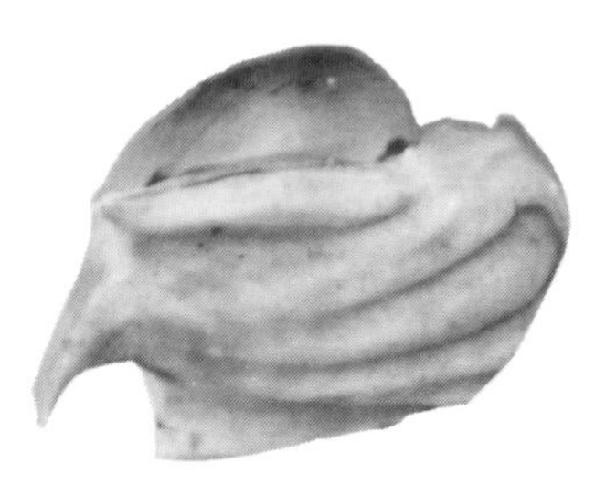

W. *Cavolinia uncinata*, 6.5 mm (Caribbean)

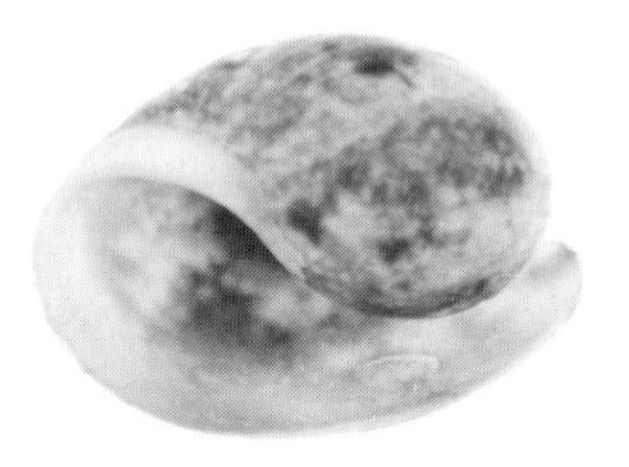

X. *Bulla ampulla*, 31 mm (Indian Ocean)

PULMONATES

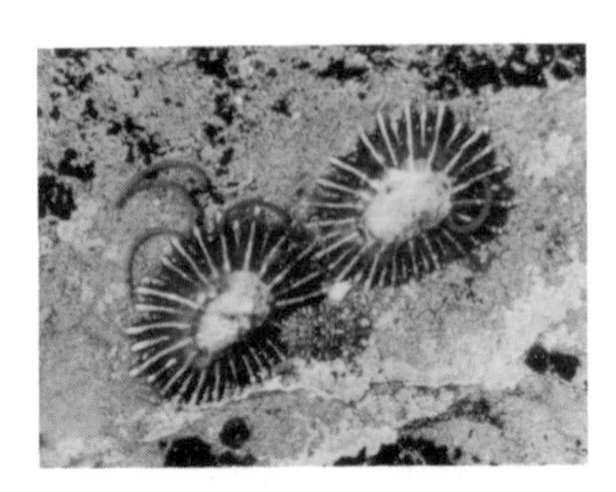

Y. *Siphonaria capensis*, 20 mm (Cape, South Africa)

Z. *Onchidium*, sp., 50 mm (Indian Ocean)

Neogastropods have a well-developed proboscis, an elaborate osphradium (olfactory organ), and a notch (siphonal canal) in the anterior lip of the shell that accommodates a snout-like fold of the mantle directing water currents on to the osphradium (section 1.7). These latter features, however, are also possessed by certain predatory mesogastropods. Neogastropods possess a haploid set of 28 to 36 chromosomes that may have arisen by polyploidy, whereas archaeogastropods and mesogastropods have haploid sets of 7 to 21 chromosomes (Patterson, 1969).

How gastropods meet their biological requirements strongly reflects their characteristic anatomical and physiological properties. These properties preclude some activities, for example sustained rapid locomotion, but are suited to others, such as close grazing of the substratum, and are remarkably adaptable as shown by the wide range of environmental conditions in which gastropods live. To serve as a point of reference for later chapters, a simple description of basic anatomical, physiological and life-historical properties of marine gastropods is given below. Comprehensive treatments may be found in Fretter and Graham (1962), Thompson (1976), Thompson and Brown (1984), Wilbur (1983, 1984), and Purchon (1977).

1.2 The Shell

Gastropods typically bear an external shell, but this may secondarily have become internal or lost altogether. Sometimes, the shell is geometrically simple, as in limpets, but in most shelled gastropods it is coiled, economically packaging the viscera. Gastropods with coiled shells are popularly known as snails and those with an internal or no shell as slugs. Secreted by the mantle (pallium), a skirt-like flap of the outer body wall (Figure 1.2), the shell consists of layers of crystalline calcium carbonate separated by thin sheets of protein. The layers are themselves composed of subunits whose orientation differs among layers, producing a 'crossed-lamellar' structure (Currey and Kohn, 1976). The organic matrix and crossed-lamellar structure help to interrupt cleavage planes and so limit the extent of damage inflicted by impact or external pressure. The inner surface is microscopically roughened in many snails, providing a grip for the mantle (Gainey and Wise, 1976).

Snails are attached to the central pillar (columella) of their shell by a strap-like columellar muscle, whose contraction withdraws the body entirely within the shell. In addition to longitudinal fibres, the columellar muscle also contains dorsoventral and transverse fibres, whose contraction elongates the muscle, tending to push the snail out of the shell (Brown and Trueman, 1982). Usually, the postero-dorsal region of the foot bears a horny or calcareous plate (operculum) which fits like a door across the shell aperture after the snail has withdrawn.

Shells are often mechanically strong, protecting marine snails from abrasion by sand, impact of moving stones, or predators. They are also impermeable barriers, protecting intertidal and estuarine snails from desiccation and fluctuating salinity. When, as among opisthobranchs, the shell is internal or lost, the mantle is expanded over the body or replaced by modifications of the dorsal epithelium, affording protection by the secretion of mucus, sulphuric acid, or toxins. Loss of the shell avoids its rigidity of form, extra mass and the metabolic cost of its secretion (section 4.6).

1.3 The Odontophore and Radula

The body of a gastropod consists of a large muscular foot, mounted on top of which are the head and visceral mass (Figure 1.2). The head usually bears a pair of tentacles, each with an eye some distance from the tip, and a short snout or retractable proboscis terminating in the mouth. The mouth opens into the buccal cavity, often equipped with one or more hard, plate-like jaws. On the floor of the buccal cavity arises a large, muscular, cartilaginous block (odontophore). Running medially over the odontophore is the radula, a flexible, chitinous ribbon bearing transverse rows of teeth, secreted from a pocket behind the odontophore. The odontophore provides the mechanical support and muscular control of the radula, which can be protruded slightly out of the mouth and used to gather food (Figure 2.1).

1.4 The Stomach and Digestive Gland

The head merges via a short, ill-defined neck into the visceral mass, containing the stomach, digestive gland, intestine, kidney, heart and gonad. Projecting into the stomach of most archaeogastropods and many microphagous mesogastropods is a stiff

Figure 1.2: A. External Anatomy of Prosobranch Gastropod. a=Anus, ct=ctenidium (gill), e=eye f=foot, m=mantle, mc=mantle cavity, op=operculum, p=proboscis (snout), s=shell, t=tentacle, vm=visceral mass. B. Alimentary Canal. a=Anus, bc=buccal cavity, i=intestine, o=oesophagus, r=rectum, st=stomach

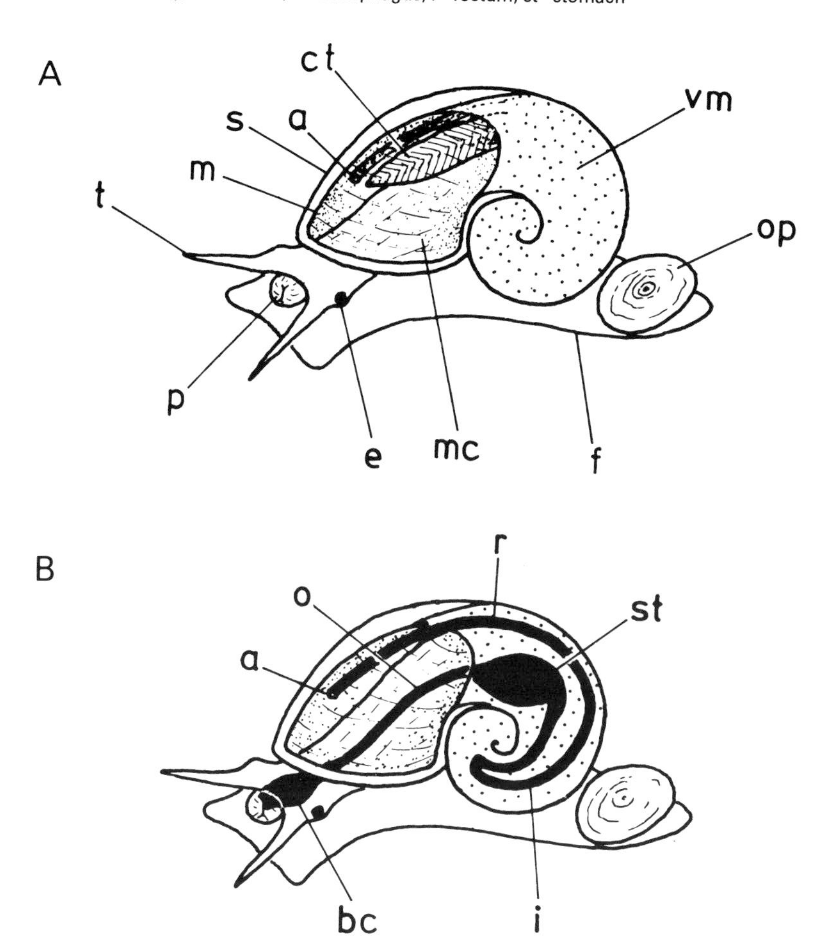

mucoprotein rod (crystalline style). The style rotates and dissolves at the tip in response to the lower pH of the stomach, liberating amylases that partially digest the food. Enzymes are also secreted into the stomach from the digestive gland (hepatopancreas), which has two characteristic types of cell: basophilic secretory cells and heterophagic digestive cells (section 3.1.2). The stomach is lined with ciliated ridges that sort particles according to size, the smallest being transported into the fine ducts of the digestive gland, the rest being bound in mucus and passed as faecal pellets along the intestine. Small particles taken into the digestive gland are phagocytosed and digested intracellularly. The majority of gastropods, however, lack the crystalline style and have a simple stomach in which food is partially digested extracellularly by enzymes secreted by the digestive gland. Fine particles of the partially digested food are taken up by the digestive gland cells within which the final stages of digestion are completed. The relative importance of extra- and intracellular digestion differs among gastropods, some, such as *Ilyanassa obsoleta*, relying almost entirely on extracellular, and others, such as *Hydrobia totteni*, relying almost entirely on intracellular digestion (Levington and Bianchi, 1981b).

1.5 The Blood System and Kidney

Metabolites dissolved in the blood (haemolymph) are circulated around the body in a system of arteries, veins, and sinuses (haemocoels) (Figure 1.3). Muscular movements during locomotion and feeding contribute to the circulation, but the haemolymph is also pumped by the heart, especially to the gills and kidney. In some marine gastropods, as in freshwater and terrestrial forms, the heart

Figure 1.3: Blood System of Gastropod. Unshaded vessels are arteries, shaded ones are veins. b—Branchial vessels, h—heart, k—kidney vessels. After Brown (1982)

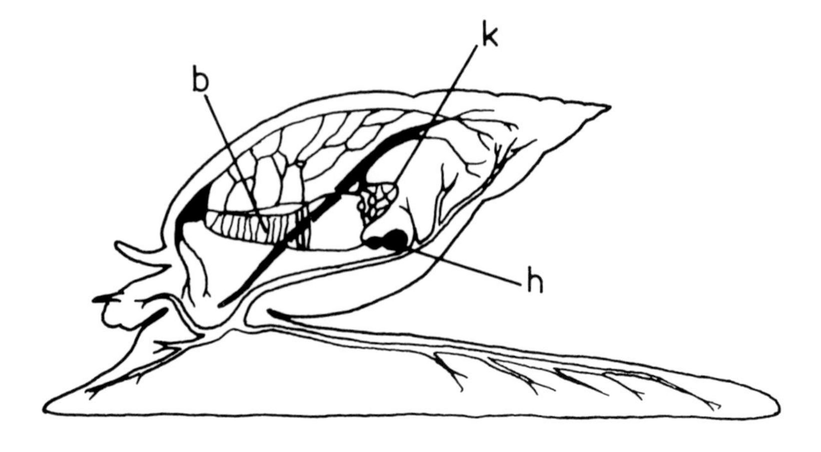

produces primary urine by ultrafiltration. When the auricle relaxes, filtration occurs across its thin wall by suction as the hydrostatic pressure in the surrounding pericardial cavity falls as a result of contraction of the ventricle. Filtration may similarly occur across the ventricle when it relaxes, but in species such as the whelk, *Buccinum undatum*, the ventricular wall is too thick for ultrafiltration (Andrews, 1976). Primary urine (pericardial fluid) filtered by the heart passes directly to the kidney where it receives the products of nitrogenous excretion. In other marine gastropods, such as *Littorina littorea*, ultrafiltration occurs not across the heart wall, but across folds of the kidney wall where nephrocytes (Figure 1.4) form an excretory epithelium producing and modifying the urine.

Figure 1.4: Renal Fold of *Littorina littorea*. Nephrocytes are separated from a renal blood vessel (bv) by a blood sinus (bs) into which they send basal extensions. The blood vessel consists of an incomplete wall of muscle (shaded) lacking an endothelial lining, so allowing blood and amoebocytes to pass outwards into the blood sinus. Fluid and solid excretions derived from the blood appear as vacuoles in the nephrocytes and are discharged into the kidney lumen (lu) by apocrine secretion. Ciliated cells among the nephrocytes aid in circulation of the urine in the lumen. After Mason and Nott (1980)

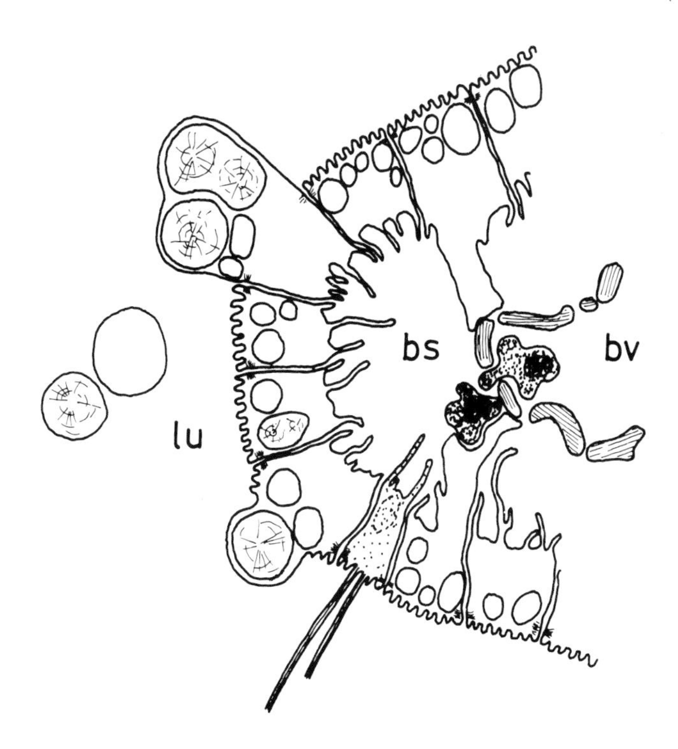

Ions are actively transported across the pericardial wall, where secretory cells form the pericardial glands, and across the kidney epithelium. Most marine gastropods, however, are osmoconformers, having body fluids isotonic with normal seawater, and have only a slight capacity to regulate ionic concentrations in the haemolymph.

In most prosobranchs, oxygen is carried by molecules of the copper-associated pigment haemocyanin, dispersed in the plasma of the haemolymph. Haemocyanin yields up to 90 per cent of its associated oxygen as it passes through the tissues, the amount carried depending on ambient oxygen tension, pH, temperature and salinity (Ghiretti, 1966; Bonaventura and Bonaventura, 1983). Molecular aggregates of haemocyanin are cylindrical, appearing round viewed end on and square viewed from the side in electron micrographs (Figure 1.5). The pigments haemoglobin and myoglobin are confined to muscles engaged in prolonged activity, particularly those of the odontophore, requiring very efficient oxygen transportation. Though it functions less efficiently than haemoglobin, haemocyanin saturates at low oxygen tensions and enables some gastropods to survive poor aeration; however, it is not present in opisthobranchs.

Gaseous exchange occurs over exposed surfaces of the body and across the gill, their relative importance depending on the size and anatomy of the gastropod. Because of its small size, less than 1.5 mm long, the prosobranch *Cima minima* has a surface area to volume ratio so large that neither a gill nor a heart is necessary to aid gaseous exchange (Graham, 1982). The moderately large snail *Bullia digitalis* has a wide, flat foot, which when expanded absorbs sufficient oxygen to meet most of the snail's metabolic requirements. The gill and heart supply oxygen to the visceral mass when

Figure 1.5: Molecular Aggregates of Haemocyanin from the Blood of *Littorina littorea*. The aggregates are cylindrical, appearing circular when viewed end on and rectangular from the side. They are 31 nm in diameter and 32 nm long. After Mason and Nott (1980)

muscular contractions impede flow through the haemocoel (Brown, 1982). In other gastropods the gill may be more important for gaseous exchange.

Gastropods without external shells respire through the external epithelium, whose surface area may be extended by accessory projections. Most naked gastropods could not withstand the desiccation associated with exposed intertidal surfaces, but the high intertidal Onchidiidae have a tough mantle that reduces evaporation and they respire through a posterior pneumostome, opening into a small vascularised mantle cavity serving as a lung.

1.6 The Mantle Cavity

The gill of shelled gastropods projects into the mantle cavity (Figure 1.2). (Mesogastropods, neogastropods and some opisthobranchs have one gill; most archaeogastropods and many opisthobranchs have more than one gill.) When exposed to the air, intertidal gastropods retain seawater in the mantle cavity. The mantle-cavity fluid prevents the gill from collapsing and is therefore crucial for efficient gaseous exchange, as shown by a reduced rate of oxygen consumption when the mantle is experimentally drained (Houlihan *et al.*, 1981). Lateral cilia on the gill filaments drive the respiratory current in the opposite direction to the flow of blood within the filaments, operating as a counter-current system for gaseous diffusion. A counter-current system probably also operates in the richly vascularised mantle rim of acmeid limpets, where cilia drive pallial water currents in a direction opposite to the flow of blood. The circumpallial network of acmeids is used mainly for aerial respiration and the gill mainly for aquatic respiration (Kingston, 1968). Certain gastropods living among the branches of mangroves experience over 95 per cent aerial exposure and have vascularised mantle cavities that function as lungs (Houlihan, 1979).

Mantle-cavity fluid, amounting to about 39 per cent of the total water content in the case of attached *Nucella lapillus* (Boyle *et al.*, 1979), may also serve as a reservoir for soluble excretory products during tidal emersion and, by evaporative cooling, may reduce thermal stress. Evaporation is confined to the mantle-cavity fluid, so protecting the tissues from desiccation and osmotic stress.

1.7 The Pallial Organs

In most prosobranchs the respiratory current generated by the gill enters the mantle cavity to the left of the head and leaves on the right. Next to the gill is a pleated, chemosensory strip of mantle (osphradium) strategically placed to detect odours in the inhalant current (Figure 1.6). The rectum discharges faeces to the right of the head in the exhalant current, with minimal risk of fouling the mantle cavity. Because of its critical role in respiration, chemoreception and waste disposal, the mantle cavity must be kept free of extraneous material. A broad strip of mucus-secreting epithelium (hypobranchial gland) runs parallel to the gill and osphradium. Detrital particles brought by the inhalant current adhere to the mucous secretions of the hypobranchial gland and are transported by ciliary tracts across the mantle cavity to the exhalant region where they are ejected as pseudofaeces.

The anatomy of the mantle cavity and pallial organs has undergone extensive changes during the evolution of prosobranchs, opisthobranchs and pulmonates (Morton, 1967). Brought from a posterior to an anterior position by torsion, an ontogenetic twisting of the body through 180° that demarcates the evolutionary origin of the gastropods (section 1.10), the mantle cavity originally had bilateral symmetry with paired gills and hypobranchial glands and was associated with paired auricles and kidneys. Coiling of the viscera left little room for sets of organs on the inner side of the spiral (usually the right-hand side), so these became suppressed, leaving their peripheral partners on the left-hand side to assume total functional capacity. No living gastropod retains perfect bilateral symmetry of the pallial system, but the archaeogastropod snail *Pleurotomaria* has paired subequal gills and hypobranchial glands. Archaeogastropod limpets have secondarily developed a non-spiral shell (although their embryonic shell is coiled) and some, such as *Fissurella*, have resumed almost complete bilateral symmetry of the mantle cavity. Others, such as *Patella*, have extended the mantle cavity right round their discoidal bodies, and the true gill has been lost in favour of numerous flaps of skin (pallial gills) that hang from the roof of the mantle cavity and enable a respiratory current to cross the entire pallial circumference. Archaeogastropod snails such as *Trochus* and *Nerita* have a single gill, hypobranchial gland and auricle, foreshadowing the condition in more advanced gastropods.

Figure 1.6: Pallial organs of Prosobranch. A. The mantle (m) has been cut and flattened out on either side. The visceral mass is not included. Arrows indicate the inhalant (left) and exhalant (right) currents. bv=Efferent blood vessel taking oxygenated blood from the gill to the heart, ct=ctenidium (gill), gd=gonoduct, h=heart (not in the mantle cavity), hg=hypobranchial gland, os=osphradium, r=rectum. B. Intact snail showing the natural positions of the pallial organs. Arrows indicate water and mucociliary currents. ct=ctenidium, hg=hypobranchial gland, os=osphradium, r=rectum

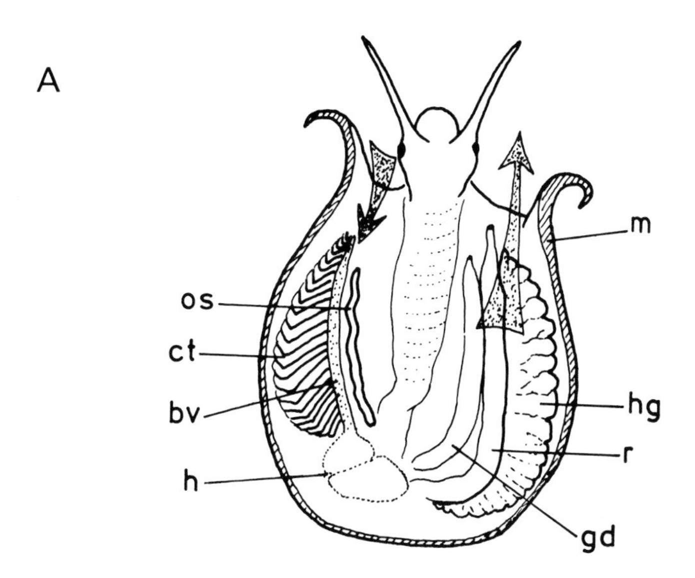

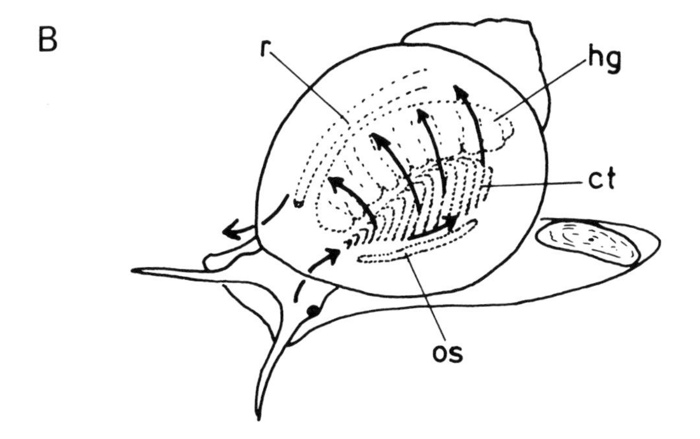

Whereas archaeogastropods retain the primitive bipectinate gill with filaments arranged on both sides of a central axis, the meso- and neogastropods have monopectinate gills with filaments only on one side of the axis. These filaments lie across the mantle cavity, directing mucus-bound particles from the inhalant (left) to the exhalant (right) side, and, together with the highly developed hypobranchial gland, form an efficient cleansing mechanism that has also become modified for particle-feeding in certain mesogastropods (sections 2.1.3 and 2.1.4).

The anterior position of the mantle cavity resulting from torsion brought the pallial organs of prosobranchs close to the head, facilitating the evolution of a powerful chemosensory mechanism that could be used for guidance to sources of olfactory stimuli. Among archaeogastropods and lower mesogastropods, olfaction is unimportant for detecting food, and the osphradium merely consists of a band of sensory cells near the base of the gill. Among predatory mesogastropods and neogastropods, the osphradium forms a bipectinate structure, superficially resembling a gill, with a greatly increased surface area. A fold of the anterior pallial margin, accommodated in a spout-like groove in the shell lip, forms a siphon that directs the inhalant current over the osphradium. In some snails, such as *Ilyanassa obsoleta*, the siphon can be extended into a long, flexible hose suitable for the precise exploration of olfactory stimuli. Heteropods, which are entirely planktonic mesogastropods (section 2.1.5), rely on their acute vision to locate passing prey and the osphradium has become vestigial.

Among opisthobranchs, the mantle cavity has tended to shift along the right side of the body towards the posterior, becoming reduced in size or even disappearing altogether. Reduction of the true gill has been compensated by dorsal projections of the body (cerata) and by the development of external secondary gills in nudibranchs or by the development of numerous parallel folds of the mantle wall to form 'plicate' gills in bullomorphs, aplysiomorphs and pleurobranchiomorphs. The planktonic thecosomes are exceptional in having a large mantle cavity which, although lacking a gill, bears a large hypobranchial gland associated with mucociliary filter-feeding (section 2.1.4).

Pulmonates have retained the anterior position of the mantle cavity but the pallial organs have been lost and the pallial wall has become highly vascularised to serve as a lung.

1.8 Mucus, the Foot and Locomotion

Slugs and snails are popularly associated with slime, and indeed external mucus is crucially important in the lives of gastropods, being used for the capture of food (section 2.1.4), repelling predators, reducing desiccation (Wolcott, 1973), avoiding freezing (Hargens and Shabica, 1973), as a buoyancy device for aquatic transportation (Vahl, 1983), for adhesion to and locomotion over the substratum, and for producing directional trails (section 4.5).

When moving over a substratum, gastropods secrete a layer of mucus from the sole of the foot, especially from its leading edge (Figure 1.7). The mucus affords a temporary grip either for gliding by the movement of cilia covering the sole or by waves of muscular contraction (Miller, 1974). Gliding by ciliary action over a mucous trail is common among species living on sediments, for example the small mud-snail *Hydrobia ulvae* or the large helmet shell *Cassis tuberosa.* Gastropods requiring a firm grip on the substratum, particularly rocky intertidal species, use muscular waves.

Elastic energy makes pedal mucus tacky so that it sticks to the substratum. When a limpet crawls, water is trapped under each new locomotory wave of the pedal muscles, serving to release the mucus from the substratum as the animal progresses. To adhere strongly to the substratum the limpet quickly expels water within the locomotory wave posteriorly, so that the whole mucous layer is immediately brought against the substratum. Adhesion is achieved solely by the tackiness of the pedal mucus, but depends on the pedal muscles being kept taut. Muscular tonus increases with the change from aquatic to aerial respiration, making limpets cling harder to rocks during low tide (Grenon and Walker, 1981).

Bullomorph opisthobranchs secrete a tube of mucus through which they move by ciliary action when ploughing through sediments or moving over hard substrata. Particles adhering to the mucus provide camouflage and the mucous tube also serves as a lanyard if the slugs drop from ledges at low tide (Rudman, 1971).

1.9 The Nervous System and Sense Organs

In primitive archaeogastropods, such as *Trochus,* the nervous system consists of a nerve ring encircling the oesophagus, a visceral loop and a pair of pedal cords. Ganglia are poorly developed, but

Figure 1.7: Pedal Muscular Waves used by Gastropods to Crawl Forwards. Second and third rows show the progression of muscular waves. A. *Neritina reclivata* uses monotaxic retrograde waves that pull the snail along by successive elongation and compression of regions of the foot. B. *Thais rustica* uses alternate ditaxic direct waves that push the snail by compression then elongation of pedal tissues, opposite to the method of *N. reclivata*. ab=Accessory boring organ, pm=pedal mucous gland. After Gainey (1976)

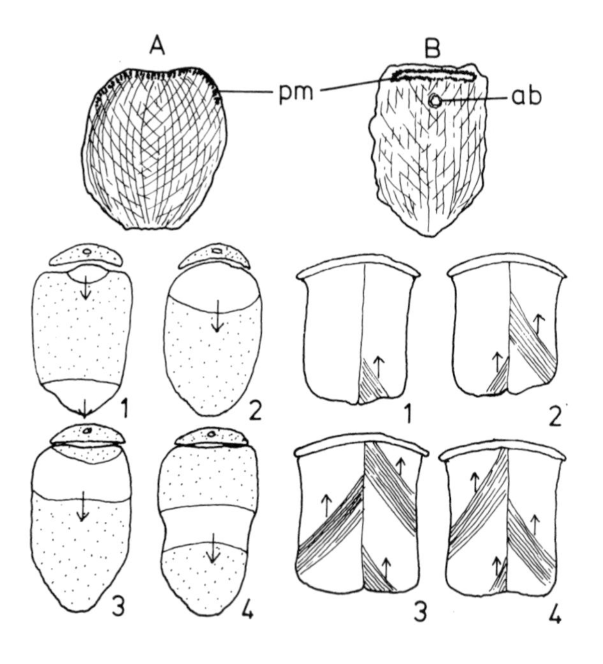

in more advanced gastropods there is an increasing tendency to group neuronal cell bodies into ganglia and for these ganglia to become clustered around the nerve ring. Lower mesogastropods, such as *Littorina*, have paired cerebral (dorsal), pleural (lateral) and pedal (ventral) ganglia on the nerve ring (Figure 1.8). The visceral loop connects supra- and sub-oesophageal parietal ganglia and paired visceral ganglia to the pleural ganglia, and is twisted into a figure-of-eight (streptoneury) by torsion during early ontogeny. Connected to the cerebral ganglia is a pair of buccal ganglia that innervate the buccal mass.

The cerebral ganglia innervate the head and control the buccal ganglia (section 2.4) and other ganglia in the nerve ring (section 7.4.2). They are also the site of neurohormonal activity influencing growth and reproduction (section 6.7). The pleural ganglia may

Figure 1.8: Nervous System of Prosobranch Gastropod. Most ganglia are grouped in a ring around the oesophagus. The visceral ganglia are connected to the others by a 'visceral loop' that is twisted round the oesophagus during the ontogenetic process of torsion. bg=Buccal ganglion, cg=cerebral ganglion, og=osphradial ganglion, pg=pedal ganglion, pl=pleural ganglion, sb=sub-oesophageal ganglion, su=supra-oesophageal ganglion, vg=visceral ganglion. All except the osphradial ganglion are paired (the sub- and supra-oesophageal ganglia form a pair)

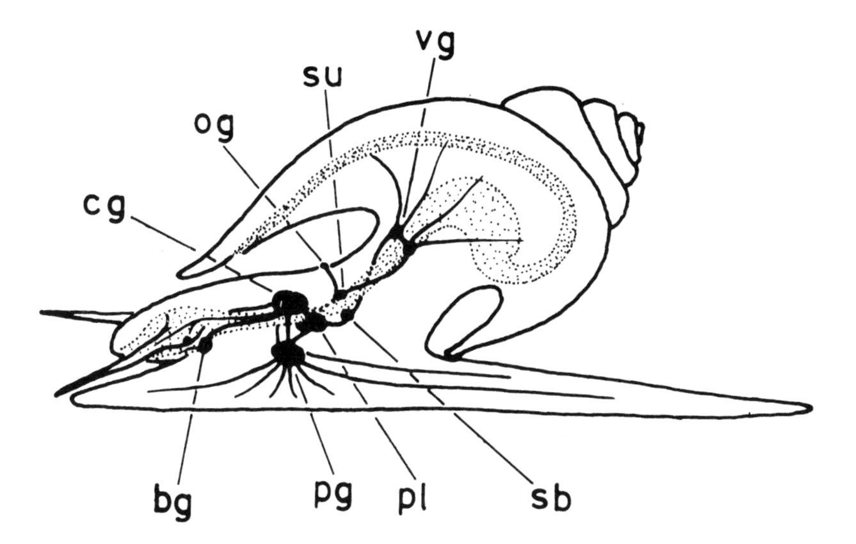

control the lateral parts of the mantle and perhaps also influence the pedal ganglia, which control the foot. The supra-oesophageal parietal ganglion innervates the left side of the mantle cavity, including the gill and, via a small peripheral ganglion, the osphradium. The sub-oesophageal parietal ganglion innervates the rest of the mantle cavity. The visceral ganglia innervate the visceral mass, including the digestive gland, stomach, intestine and reproductive tract.

In predatory mesogastropods, neogastropods, opisthobranchs and pulmonates, all principal ganglia are centralised about the nerve ring and the figure-of-eight configuration of the visceral loop is lost. The nerve ring of opisthobranchs is readily accessible by dissection and contains unusually large cell bodies, facilitating the placement of microelectrodes for recording electrical activity of individual neurons. Large opisthobranchs, such as *Aplysia* and *Tritonia*, are therefore popular subjects for neurophysiological experiments (sections 2.4 and 7.4.2).

Gastropods are equipped with organs sensitive to touch, smell, gravity and light. Whereas probably none is sensitive to airborne sound, some gastropods can detect low-frequency vibrations in the substratum. For example, *Polinices incei* can detect vibrations generated by the burrowing activities of bivalves upon which it feeds, and the snail can be made to respond to emissions from loudspeakers placed on the sand (Kitching and Pearson, 1981).

The cephalic tentacles on either side of the head (Figure 1.2) are sensitive to touch and smell. Some gastropods, such as *Monodonta* and *Patella*, have additional pallial tentacles around the mantle rim. The cephalic and pallial tentacles are often covered with immobile cilia, or compound cirri, connected to innervated sensory cells. A special chemosensory organ, the osphradium, is present in the mantle cavity and detects odours borne by the inhalant respiratory current (section 1.7). The osphradium is reduced or lost in pulmonates and in many opisthobranchs.

Simple statocysts (small cavities lined with ciliated cells in contact with limestone spherules) are embedded in the foot near to the pedal ganglia and allow orientation with respect to gravity.

Each cephalic tentacle bears an eye, which shows varying degrees of complexity according to the phylogeny and habit of the gastropod (Fretter and Graham, 1962). In archaeogastropod limpets, the eye is a simple cup-shaped pit (optic vesicle) lined by a retina comprised of photoreceptive cells shielded behind from stray light by a layer of densely pigmented cells. In other archaeogastropods, such as *Trochus* and *Nerita*, the optic vesicle is almost closed off from the exterior and is filled with an aqueous 'humour'. In mesogastropods, neogastropods, opisthobranchs and pulmonates, the optic vesicle contains a spherical lens and is closed by a transparent cornea. The pelagic heteropods have tubular telescopic eyes with a large lens and a greatly elaborated retinal surface, giving these predators exceptionally acute vision (section 2.1.5). Burrowing gastropods, on the other hand, tend to have secondarily reduced eyes sunk beneath the skin or may even have lost them altogether.

1.10 Reproduction and Life History

So far as is known, all marine gastropods proliferate by sexual reproduction (Chapter 6). Fertilised eggs develop into ciliated veliger larvae, each bearing a delicate shell (protoconch). During its development, the veliger is twisted about the neck through 180° (torsion) by contraction of an asymmetrical retractor muscle attached to the protoconch and subsequently reinforced by asymmetrical growth of the larva. Torsion brings the mantle cavity and pallial organs into a dorsal, forwardly directed position, hence the name Prosobranchia, meaning 'forward-gill'. The Opisthobranchia ('hind-gill') have become reorganised towards bilateral symmetry, losing all vestiges of torsion except for a transient stage of embryological development.

Primitively, as among archaeogastropod limpets, eggs are fertilised externally, developing into veligers that feed in the plankton, finally settling on a suitable substratum where they metamorphose into juveniles with most of the anatomical features of the adult. This basic life history has become variously modified among gastropods. Most mesogastropods, neogastropods and opisthobranchs fertilise eggs internally by copulation and lay them in protective jelly or capsules. Veligers often metamorphose within the capsules, or sometimes within the mother's reproductive tract, emerging as crawling juveniles without negotiating a planktonic phase (section 6.6).

2 ACQUISITION OF FOOD

2.1 Methods of Feeding

2.1.1 Introduction

Most gastropods superficially look similar, recognisable as 'snails' or 'slugs', yet their basic anatomy has facilitated a tremendous radiation of feeding methods, perhaps paralleled only by the Crustacea. The radula is an evolutionary versatile apparatus, ranging from rasps suitable for scraping algae off rocks, or brushes for sweeping up fine particles, through zippers with sharp teeth proficient at tearing and grasping flesh, to armaments of barbed teeth capable of harpooning prey, or delicate stylets that can lance the individual cell walls of algae. Mucus may be secreted to form sticky nets or fine meshes for capturing suspended food particles. These attributes, together with the ability of some species to incorporate chloroplasts in their tissues and the modification of others for parasitism, enable gastropods to occupy nearly every possible trophic category (Kohn, 1983).

2.1.2 Grazers and Macro-herbivores

Archaeogastropods and many mesogastropods have rasp-like radulae used for scraping food off the substratum, probably as did ancestral gastropods. Limpets, littorines, nerites and trochids exemplify grazers of the micro-algae growing on rocks, shells or on larger plants. When feeding, the radula is pulled outwards, flexing over the apex of the odontophore (Figure 2.1) and causing the teeth in every row to spread out and become erect. The odontophore is protruded slightly and rocked forwards so that the radular teeth are scraped against the substratum. After each stroke, the lips close over the retracting odontophore and the radula slides backwards, pulling in the food. In *Patella vulgata*, and perhaps also in other grazers, a plate-like jaw is held close to the retracting radula, preventing the loss of food material (Runham and Thornton, 1967).

Scraping wears down the teeth at a rate dependent on the hardness of the substratum (Figure 2.2). *Patella vulgata* grazes the surface of rocks and its teeth are hardened with silica and ferric

Figure 2.1: Movement of the Prosobranch Odontophore and Radula during Grazing. A. Fully retracted odontophore and radula. B. Fully protracted odontophore. The radula has been pulled to its fullest extent around the odontophore. C. Completed grazing stroke. The odontophore has been pivoted forwards and the radula partially retracted. bc=Buccal cavity, j=jaw o=oesophagus, od=odontophore, op=odontophore protractor muscle, or=odontophore retractor muscle, ra=radula, rp=radula protractor muscle, rr=radula retractor muscle, rs=radula sac. Only certain muscles are shown. The odontophore and its muscular complex constitute the buccal mass

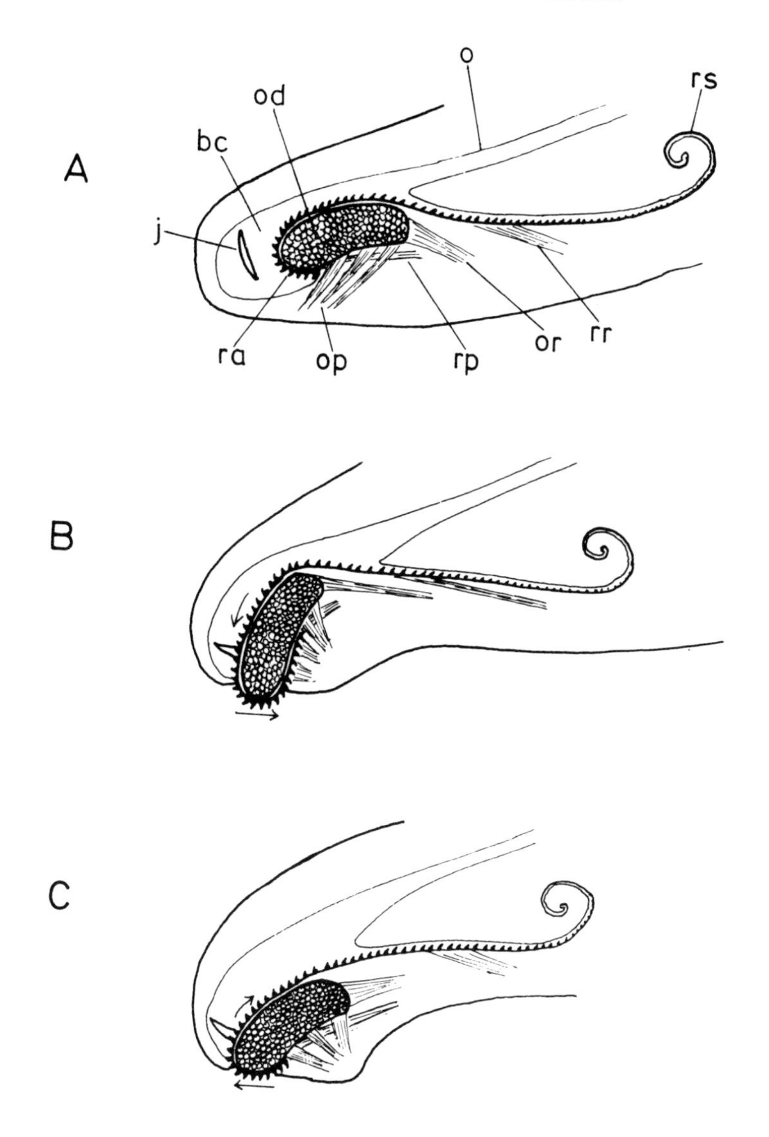

Figure 2.2: A Wearing of the Central (Rachidian) Tooth of the Radula of *Littorina littorea* (Lateral and Marginal teeth not shown). When snails are fed for 2 weeks on the delicate alga *Ulva lactuca* (left), the tooth shows no signs of wear, but when fed on the tougher algae *Chondrus crispus, Hildenbrandia rubra* and *Ralfsia expansa,* the teeth become obviously worn. B. Toughness of the algae measured as the weight required to pierce the plant with a flat-tipped 3 mm steel rod. After Bertness *et al.* (1983)

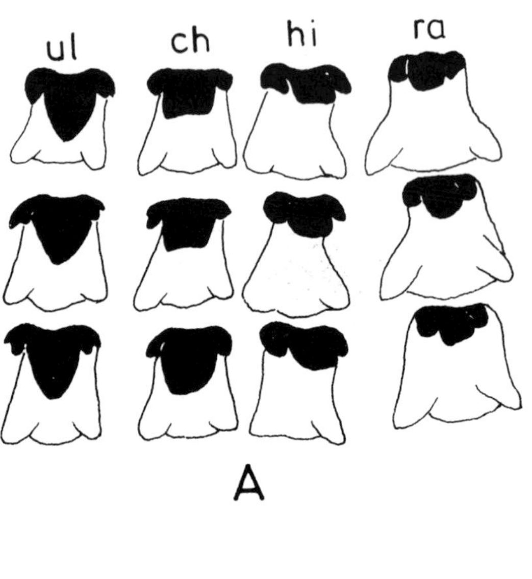

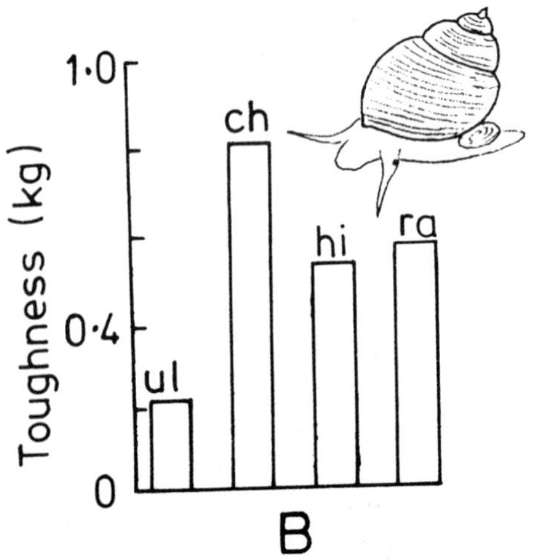

oxide (Runham *et al.*, 1969). The front of the cusps is rich in iron, the back in silica, and fibres in the protein-chitin complex forming the substance of the teeth run normal to the substratum in the leading edge of cusps and parallel on the backface. As a result of this construction, differential wear causes the teeth to function as self-sharpening chisels (Figure 2.3). The teeth have a hardness of about 5 on Mho's scale, comparable with softer rocks, but they quickly wear out. Consequently the radula of *P. vulgata* is long and secreted rapidly, producing several rows of teeth per day (Runham and Isarankura, 1966). Not all grazing gastropods are so robust. The Caecidae, for example, are part of the meiofauna living among the interstices of sandy gravel. They have delicate teeth (Figure 2.4D) which, rather than scraping hard against the substratum, are suitable for brushing individual diatoms off sand grains.

Grazing gastropods can therefore be placed in functional groups based on their feeding methods and radular morphology (Figure 2.4). Non-limpet archaeogastropods have relatively delicate teeth

Figure 2.3: Radular Tooth of *Patella vulgata*. Organic fibres run almost normal to the substratum at the leading edge and parallel at the back edge. The fibres direct the orientation of crystalline growth of minerals, causing differential wear that keeps the tooth sharp. Minerals at the leading edge contain 47% iron and 3% silicon compared with 26% and 30% at the rear. After Runham *et al.* (1969)

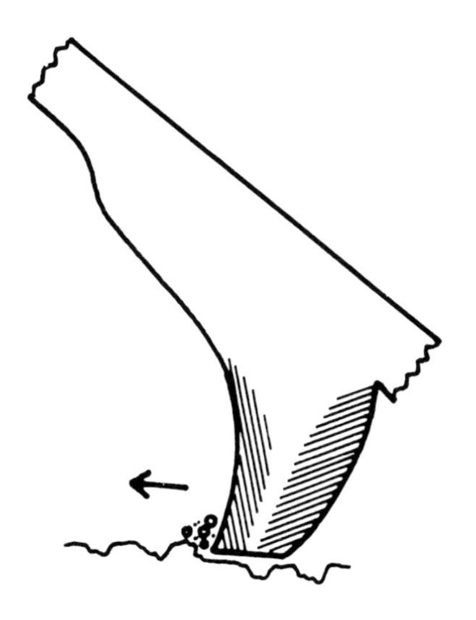

Figure 2.4: Radular Morphology of Grazing and Deposit-feeding Gastropods. (A) Non-limpet archaeogastropods, (B) mesogastropods, (C) patellacean limpets. In these 'functional groups' the effective width of the grazing-stroke and the number of functional tooth-points decrease whereas the excavation capability increases from A to C. After Steneck and Watling (1982). (D) Interstitial diatom-feeder, *Caecum*. The head of the inner marginal tooth is hinged, probably allowing flexion while sweeping over sand grains to rake in diatoms. After Hughes (1985). (E) Deposit-feeder, *Hydrobia ulvae*. After Seifert (1935). (D) and (E) show only one-half of the row of teeth. Drawings not to scale.

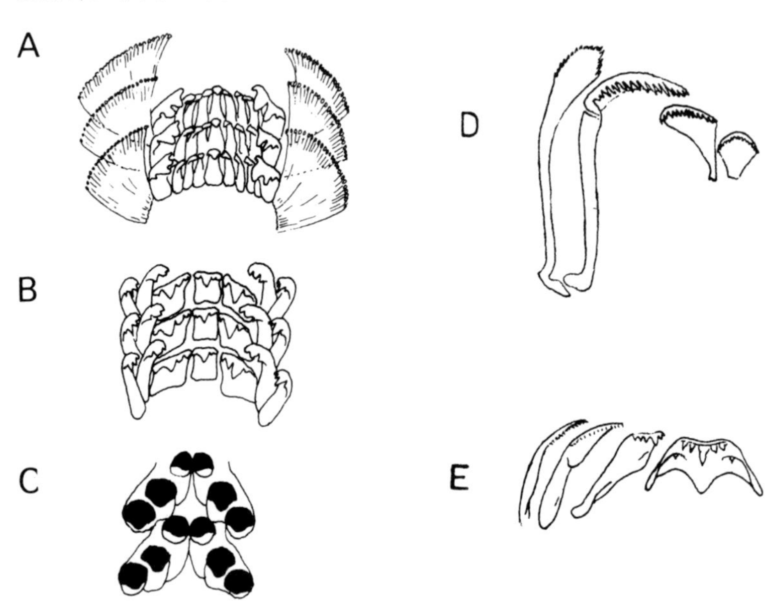

and the radula functions like a broom, sweeping the substratum broadly, but with little force. Mesogastropod radulae function like rakes capable of cropping large, tougher algae. Limpets have robust, hardened teeth with fewer points of contact on the substratum, and the radula is firmly attached to the odontophore by robust muscles enabling deeper excavation of hard surfaces as the odontophore forcefully scrapes the radula against them.

Functioning of the radula, however, may be far more complex than is apparent from its gross anatomy. Time-lapse photographic analysis of the radular stroke, coupled with microscopical examination of the radular scrape-marks on experimental surfaces, have revealed an amazingly intricate and versatile grazing mechanism in the western North American archaeogastropod *Tegula funebralis* (Hickman and Morris, 1985). Posteriorly to the odontophore, the radula is rolled into a cylinder with the teeth on the inside, but it unrolls towards its anterior end, under the control of muscles linking the radular membrane to the odontophore. As the radula slides back over the odontophore during the grazing stroke (section 1.3), the anatomically posterior rows of teeth scrape the substratum first, followed by two or more anterior rows. Because the splayed radular ribbon folds back into a cylinder as it passes the odontophore, the teeth scrape the substratum towards the midline, raking in loosened material from the sides. Only the marginal teeth scrape the substratum, as the central (rachidian) and lateral teeth are recessed in a median longitudinal fold of the radular membrane where it overlies the groove between the paired odontophoral cartilages. Circumstances under which the central and lateral teeth are used have not been determined.

Considerable flexibility in the mode of radular operation is possible because the outward marginal teeth can be mechanically linked to act in concert, whereas the odontophoral cartilages can be moved independently of each other, allowing pressure to be applied unequally and in an alternating sequence on either side of the radula. This versatility is exercised by *T. funebralis* when grazing different kinds of substrata. On the relatively flat blades of kelp, the inner marginal teeth are used to make deep incisions, releasing algal cell contents. Repetition of this grazing mode in the same location causes substantial excavation of the algal tissue. On the irregular surfaces of certain red algae the mid and outer marginal teeth are brushed over the substratum, under gentle pressure from the odontophore, sweeping up attached microbiota without penetrating the underlying algal tissue.

Mechanical interactions among teeth are particularly evident in the extensive tooth-rows of archaeogastropods (Hickman, 1984). The shafts of central and lateral teeth are generally shorter, reinforced with more complex cross-sectional shapes, and less flexible than those of marginal teeth, reflecting an excavating rather than a predominantly brushing mode of operation. Tooth shafts often have lateral chitinous flanges that interlock among teeth in a row, and basal expansions of the tooth shafts show diverse methods of overlapping and interlocking both within rows across the radula and in columns along its length. The basal expansions of the tooth shafts have several important functions. They dissipate stress applied to the radular membrane by the teeth; they counteract bending and overturning of individual teeth; they help to align, orientate and co-ordinate teeth during the grazing stroke; and they facilitate economical folding of the teeth when the radula is retracted.

Different radular structures and movement, or similar radular action on different substrata, generate different sounds which in some circumstances can be detected with a stethoscope or with a contact microphone (Kitting, 1979). These radular scraping noises can reveal whether the gastropod is brushing microscopic material off the surface of a plant or gouging into the plant itself. When several different algal species are covered by the gastropod, grazing sounds may even enable the one being eaten to be distinguished from the others.

Whereas grazing gastropods commonly feed on microscopic material covering the surface of larger algae such as fucoids and kelps, few eat the substratum itself, perhaps because these macro-algae have developed herbivore-repellent properties. Certain gastropods that do eat large brown algae tend to be dietary specialists, for example the Californian *Acmaea incessa*, the South African *Patella compressa* and the European *Helcion pellucidus*, perhaps having evolved specific mechanisms for coping with the plants' defences. However, some generalist herbivores, such as *Tegula funebralis*, also eat brown algae. Rapidly growing, ephemeral, green algae, on the other hand, lack defences and are readily consumed by grazers, as shown by the appearance of a green sward on rocks experimentally cleared of limpets (section 7.4.4).

Aplysiomorph opisthobranchs (Figure 3.3A) are among the few gastropods equipped to deal with large pieces of plant material. They crop macro-algae with their jaws and radula, passing pieces of algal tissue into a storage crop. From there the food enters a gizzard with two chambers, the first for grinding up the food, the second lined with delicate projections for straining the pulp. The finest particles are transported to the digestive gland, the rest are defecated.

Unlike archaeogastropod limpets, in which the digestion of micro-algae is almost entirely intracellular within the digestive gland, pulmonate limpets, e.g. *Siphonaria*, process food almost entirely by extracellular enzymes in the stomach, enabling them to cope with large fragments bitten off macro-algae (Branch, 1981).

2.1.3 Deposit-feeders

The surface of sand and mud provides a habitat for numerous

micro-organisms. At depths shallow enough for the penetration of sufficient light, diatoms, blue-green and filamentous algae grow over the substratum and at all depths sedimentary particles are coated with bacteria. Both micro-algae and bacteria are more numerous in sheltered areas, where the sediments are more stable, contain more organic detritus and are finer, with a greater surface area per particle. Sands and muds therefore contain a rich supply of food that supports diverse animals. Among these are deposit-feeding snails such as *Hydrobia* spp., superabundant on intertidal mudflats, which have fine radular teeth suitable for raking in sediment (Figures 2.4E, 2.5A). *Aporrhais pes-pelecani* is a more sedentary deposit-feeder. It buries itself just beneath the substratum (Figure 2.5B) and extends its proboscis to browse on the surrounding detritus.

2.1.4 Suspension-feeders

The ctenidial and pallial muco-ciliary systems for generating the respiratory current and ridding the mantle cavity of fouling particles (section 1.6) have become modified in various gastropods

Figure 2.5: Deposit-feeding Gastropods. A. *Hydrobia ulvae* (3-6 mm) grazes the surface of the deposit. B. *Aporrhais pes-pelecani* (3-4 cm) feeds with a mobile proboscis from beneath the surface of the sediment. Having exhausted the food within reach, it emerges and moves a little further on before reburrowing. Inhalant and exhalant tunnels, constructed by the proboscis, are lined with mucus. After Yonge and Thompson (1976)

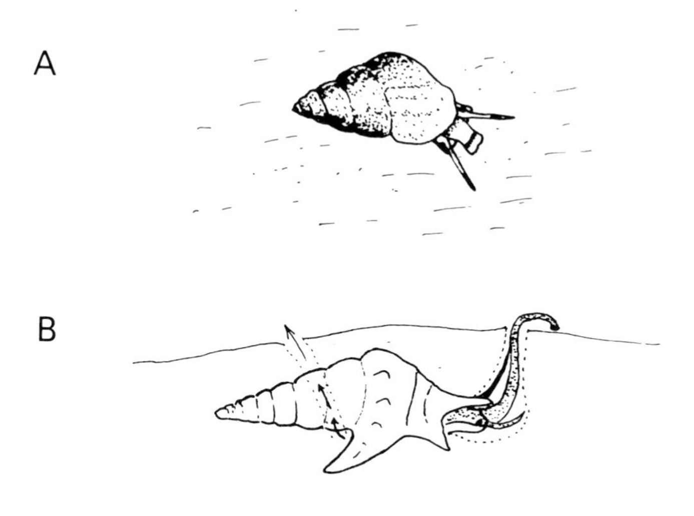

for intercepting suspended food particles. In the slipper-limpet, *Crepidula fornicata*, a coarsely meshed mucous screen secreted across the inhalant aperture (Figure 2.6A) strains off particles too large to be handled by the pallial muco-ciliary system, and smaller particles are intercepted by a finer-meshed mucous screen over the surface of the gill, which retains particles down to 1-2 μm in diameter. Particles bound in mucus are transported by ciliary tracts out of the mantle cavity, picked up by the radula and ingested.

The pulmonate limpet *Gadinalia nivea*, hanging beneath ledges on New Zealand shores, secretes a mucous bag that billows out with the surge of waves, straining off suspended particles (Figure 2.6B). A similar mechanism is used by *Olivella columellaris* to catch particles drifting in the backwash down sand beaches on the Pacific coast of Central America (Figure 2.6C).

Hydrothermal vents, occurring on the tectonically active ridges of the deep ocean floor, allow the prolific growth of sulphur bacteria that support a diverse community of filter-feeders. Among these are a number of limpets, some of them as yet unnamed. One of these species, from the North East Pacific, has the large gills typical of filter-feeding limpets, but the gills are also colonised by a dense mat of filamentous bacteria. Electron microscopy and histochemical techniques have shown that some of these bacteria are taken across the ctenidial epithelium by endocytosis (Burgh and Singla, 1984). This may form an alternative route to ingestion but other possibilities, such as a defensive reaction to infestation, cannot be ruled out by the data.

The Vermetidae, which in the post-larval stages have partially uncoiled shells cemented permanently to the substratum, secrete quantities of mucus from an enlarged pedal gland. Special pedal tentacles spin off the mucus as fine threads which are spread out by water movement over the substratum, coalescing into a web (Figure 2.6D) that catches suspended particles and even small animals such as crustacean larvae. The web is periodically hauled in by the radula and ingested. On tropical and warm temperate shores exposed to strong wave action, species of *Dendropoma* form dense aggregations in a well-defined 'vermetid' zone (Stephenson and Stephenson, 1972; Hughes, 1979a).

In more sheltered localities, turritellids in the subfamily Vermiculariinae cement their vermetid-like shells to stones. Particles brought by the inhalant current adhere to mucus and are transported along the gill filaments and along pallial ciliary tracts to a

Figure 2.6: Filter-feeding Gstropods. (A) *Crepidula fornicata* (2-3 mm long). A coarse-meshed mucous filter spans the entrance to the mantle cavity and a fine-meshed one covers the frontal surface of the gill. Particles suspended in the inhalant current are caught by the filters. Those trapped in the ctenidial mucus are transported by the frontal cilia to the edge of the gill and thence to the food-groove. After Jørgensen *et al.* (1984). (B) *Gadinalia nivea* The limpet (1.5 cm long) attaches in clusters to the roofs of caves or beneath ledges where there are strong surge currents. Water enters the mantle cavity from behind and exits anteriorly, where it passes through the billowing mucous net secreted by the adjacent mantle. The oral lobes (ol) scan the net, and if it is full of food, they gather it up for ingestion. After Walsby *et al.* (1973). (C) *Olivella columellaris*. The snail (1 cm long) is buried in the sand with the head projecting above the surface. Mucous bags are spun from the pedal 'horns' and billow out in the swash of receding waves. Periodically, the nets are drawn forwards and the proboscis reaches from behind to ingest them. After Seilacher (1959). (D) *Dendropoma corallinaceum* (1.5 cm long). Mucous threads spun from the grooved pedal tentacles are spread by currents over the substratum, forming a net. At 1-2 min intervals, the net is ingested and another secreted. After Hughes (1978). (E) *Umbonium vestiarium* (shell 1.0 cm diameter). Flaps of tissue from the upper surface of the foot form an inhalant and an exhalant siphon. The ctenidium is greatly enlarged and generates a substantial inhalant current drawing in suspended food particles. The particles are trapped in mucus secreted by the hypobranchial gland and transported to the tips of the gill filaments. The mucous-bound particles then pass along the ciliated food-grove, out of the mantle cavity to the vicinity of the snout, where they are grasped by the jaws and ingested by the radula. Filtering can take place while the snail is crawling (upper left) or when it is buried in the sand (upper right). CT=ctenidium (gill), E=eye, ES=exhalant siphon, FG=food-groove, HG=hypobranchial gland, IS=inhalant siphon, OS=osphradium, R=rectum, T=tentacle, VM=visceral mass. After Fretter (1975)

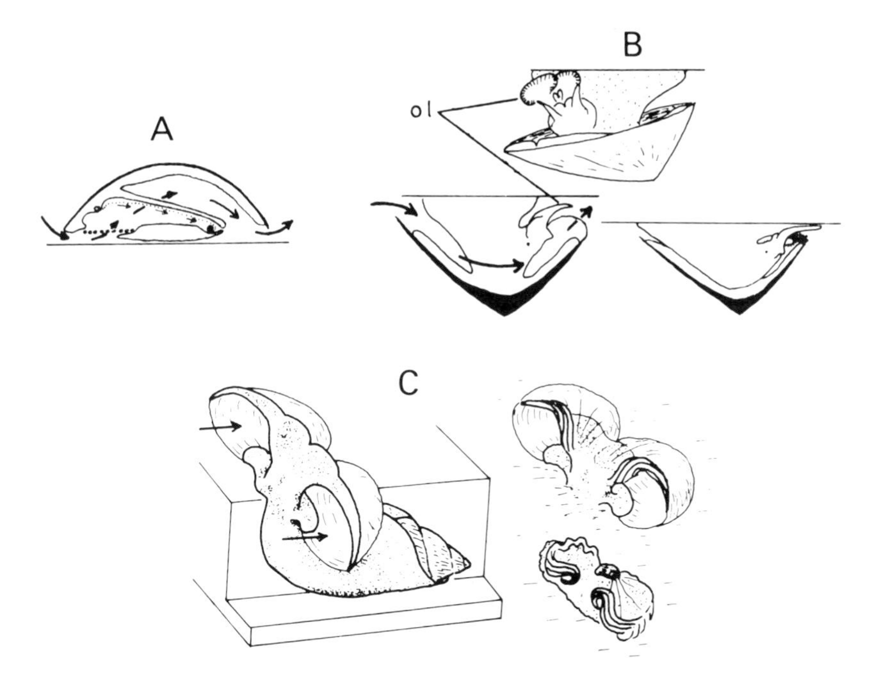

Figure 2.6 continued

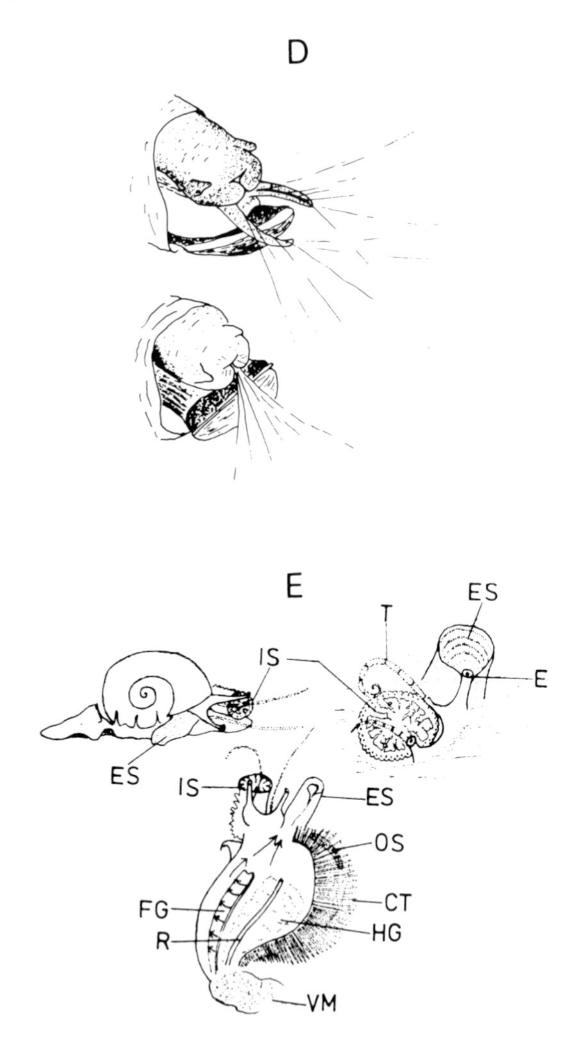

lobe protruding out of the mantle cavity. Here the food-laden mucus is periodically removed by the radula and ingested (Morton, 1951; Hughes, 1985b). A similar branchial filtering mechanism is employed by *Turritella* (Graham, 1938) and by vermetids (Hughes, 1978) in addition to their deposit-feeding activities. On tropical sandy beaches, trochid snails in the genus *Umbonium* filter particles suspended in the water at high tide or in water draining down the shore. They have similar pallial filtering mechanisms to vermiculariids and turritellids, but also have well-developed inhalant and exhalant siphons (Figure 2.6E).

Small, shelled opisthobranchs in the order Thecosomata swim near the ocean surface by flapping their lobed foot. Using muco-ciliary tracts on the floor of the mantle cavity and on the foot, thecosomes in the family Cavoliniidae intercept planktonic diatoms from the inhalant current (Gilmer, 1974), crushing their siliceous cases in the oesophageal gizzard (Figure 3.3B). Other small planktonic organisms and fine detritus are also ingested. Species such as *Gleba cordata*, in the family Cymbuliidae, secrete a mucous web, up to 2 m in diameter, from the margins of the wings. The snail hangs below the web, clinging to it with a long outstretched proboscis. Planktonic organisms and fine detritus particles are trapped by the web, and are consolidated in a mucous string secreted by the proboscis and ingested (Gilmer, 1972).

2.1.5 Carnivores

Some mesogastropods, most neogastropods and many opisthobranchs are predators, whose methods of attack vary tremendously, reflecting their own taxonomic diversity and that of their prey. Predatory prosobranchs (except the Naticacea) differ from grazers by the development of a siphonal canal in the anterior lip of the shell. The siphonal canal receives a fold of the mantle that directs the inhalant current on to the highly developed osphradium detecting the scent of prey (section 1.7). Other differences include modifications of the radula (Plate 2.1A), proboscis and associated glands for attacking prey (Ponder, 1973; Hughes and Hughes, 1981). Equally specialised modifications, including buccal pumps and oesophageal gizzards, are found among carnivorous opisthobranchs (Rudman, 1972a, b; Thompson, 1976; Todd, 1981).

Among mesogastropods, many cowries (Figure 2.7A) graze sponges, coelenterates and colonial tunicates in much the same way as other members of their superfamily (Cypraeacea) graze algae and detritus. The Tonnacea contains families of entirely carnivorous species, all of which have an extensible proboscis associated with glands that secrete sulphuric acid or toxins (Hughes and Hughes, 1981). Tonnidae feed on holothurians which they engulf within a few minutes using their greatly dilatable proboscis

Plate 2.1A: The Radular Teeth of *Nucella lapillus* Serve Two Functions. The robust central teeth excavate shell material when the snail drills through the exoskeleton of its prey. The cusps of the central teeth become worn by abrasion, as shown by the lower (anterior) teeth in the photograph. The more delicate, hooked, lateral teeth are splayed out on the fully protracted radula, but fold inwards and intermesh like a zipper when the radula retracts over the apex of the odontophore, as shown in the upper region of the photograph. This enables the lateral teeth to grasp strands of flesh firmly during ingestion of the prey. Magnification: × 80. SEM by N.W. Runham.

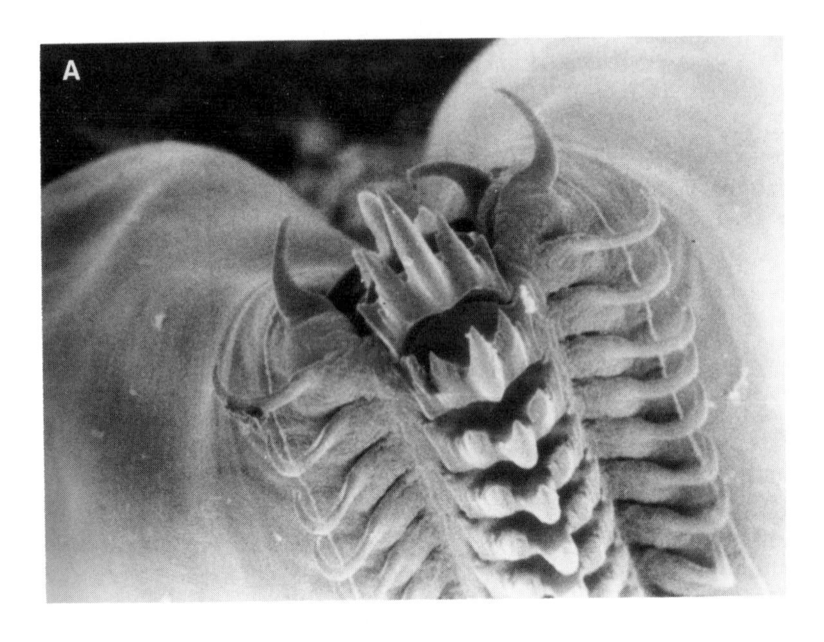

(Figure 2.7C). Cassidae feed on echinoids, dissolving the shell with sulphuric acid and cutting out a disc of the softened material within about 10 min to gain access to the inner flesh (Figure 2.7D). Bursidae and Cymatiidae feed on polychaetes, sipunculans, bivalves and ascidians which they anaesthetise with toxins (Houbrick and Fretter, 1969; Taylor, 1984). Members of the superfamily Naticacea feed on gastropods and bivalves, usually drilling a hole in the shell by alternate application of the accessory boring organ and radular scraping. The accessory boring organ opens near the front of the sole (Figure 2.8B), producing an acid secretion containing carbonic anhydrase that softens the prey's shell. Penetration is slow and may take several days with thick-shelled prey (Hughes, 1985c). Among the most spectacular

Plate 2.1B,C: Species of *Conus* Have a Highly Modified Radula. The central and lateral teeth are wholly suppressed, leaving only a pair of marginal teeth in each row. The marginal teeth are shaped like a harpoon and are grooved to carry a neurotoxin (photograph B) secreted by a large poison gland. One tooth is used at a time, being carried at the tip of the proboscis as it darts into the prey. Although tethered to the proboscis to prevent escape of the struggling prey, the tooth is finally discarded as the prey is being consumed and another tooth is 'loaded' into firing position for the next attack. In this specimen, the posterior barb is itself hooked at the tip (photograph C), assuring a secure grip when embedded in the prey. Scanning electron micrographs by N.W. Runham. Magnification: B = × 100, C = × 270

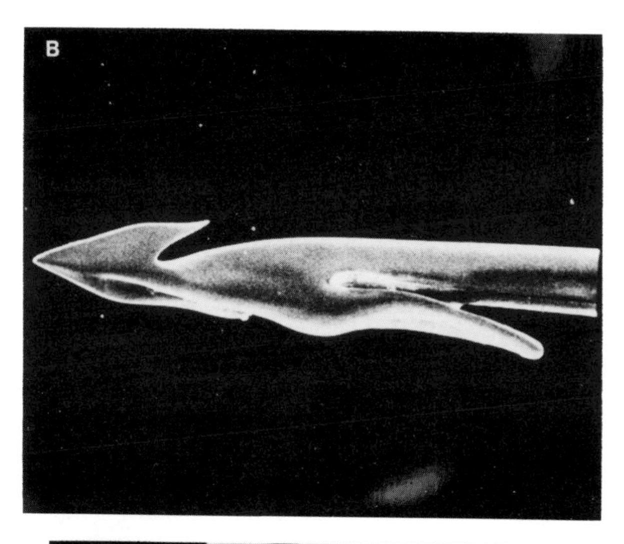

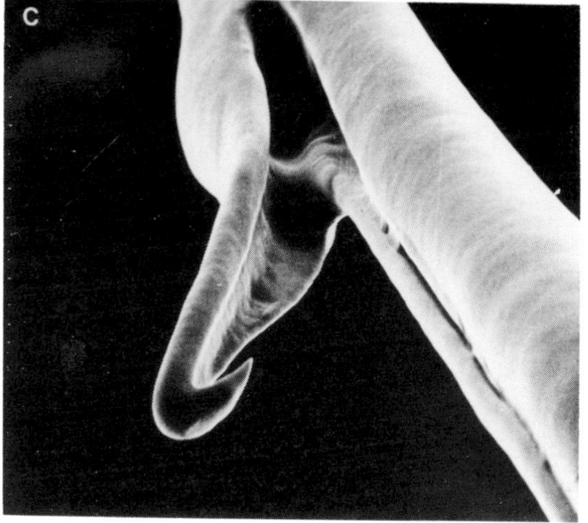

carnivorous mesogastropods are the Heteropoda, highly modified for planktonic life, with transparent bodies and muscular fins (Figure 2.7E). Equipped with large, stalked eyes and an extensible proboscis, heteropods pounce upon pteropods and other planktonic prey including small fish. *Cardiapoda placenta* feeds on pelagic tunicates and can detect the silhouette of prey swimming above it. From a distance of as much as 60 cm, this heteropod darts at a speed of up to 40 cm^{-1}, seizes the prey with its buccal cones and ingests it in about 10 min (Hamner *et al.*, 1975).

Neogastropods, probably derived from archaeogastropod or early mesogastropod ancestors independently of the carnivorous mesogastropods (Ponder, 1973), show a particularly wide range of specialised predatory characteristics, although some, such as *Olivella columellaris* (Figure 2.6C), have become secondarily non-predatory. Buccinacea are opportunistic feeders readily scavenging moribund animals and also attacking intact prey. The whelk *Buccinum undatum* is attracted to dead fish in baited pots, but stomach contents reveal a dietary predominance of polychaetes with smaller amounts of various other taxa including bivalves. On encountering a cockle, *B. undatum* positions itself at the edge of the prey's shell and waits for the latter to gape, whereupon the whelk instantly wedges the lip of its own shell between the valves of the cockle and ingests the exposed flesh (Figure 2.8A). Shells of *Fasciolaria* spp. have a thickened, sculptured lip that is brought forcefully against bivalve prey, chipping a hole through which the proboscis is inserted (Paine, 1966). Muricacea use an accessory boring organ in the foot (Figure 2.8B) to penetrate the shells of barnacles, gastropods and bivalves as described above for the Naticacea (Carriker and Van Zandt, 1972). The robust central (rachidian) radular tooth serves for excavation while the curved lateral teeth grasp the prey's flesh as they move over the odontophore (Plate 2.1A). Predatory members of the Volutacea envelop bivalves and gastropods in their large foot until the prey begins to relax, probably by asphyxiation, whereupon the proboscis is inserted to ingest the flesh. Harpidae quickly engulf crabs and other decapod crustaceans in their foot, secreting a sticky mucous envelope which immobilises the prey. Mitridae extend their proboscis to engulf sipunculan worms (Fukuyama and Nybakken, 1984). The toxoglossan families Turridae, Terebridae and Conidae have dart-like radular teeth (Plate 2.1B, C) that harpoon prey, ranging from polychaetes to small fish (Figure 2.8C), and in most cases inject toxin

Figure 2.7: Predatory Mesogastropods. A. Cypraeacean *Erato voluta* (1 cm) feeding on the ascidian *Botryllus schlosseri*. After Fretter (1951). B. Naticacean *Natica unifasciata* (3 cm) feeding on a smaller conspecific. In the left figure the predator is extending its foot over the prey. In the right figure, the prey is completely enveloped by the predator's foot. The proboscis can be seen in the drilling position. After Hughes (1985c). C. Tonnacean *Tonna perdix* (5 cm) engulfing a holothurian. D. Cassid *Cypraecassis testiculus* (6 cm) drilling a sea urchin *Diadema antillarum*. After Hughes and Hughes (1981). E. Heteropod *Carinaria mediterranea* (10 cm). The animal is depicted in its swimming position. When not swimming, pteropods curl themselves into a loose, neutrally buoyant ball. After Fretter and Graham (1962).

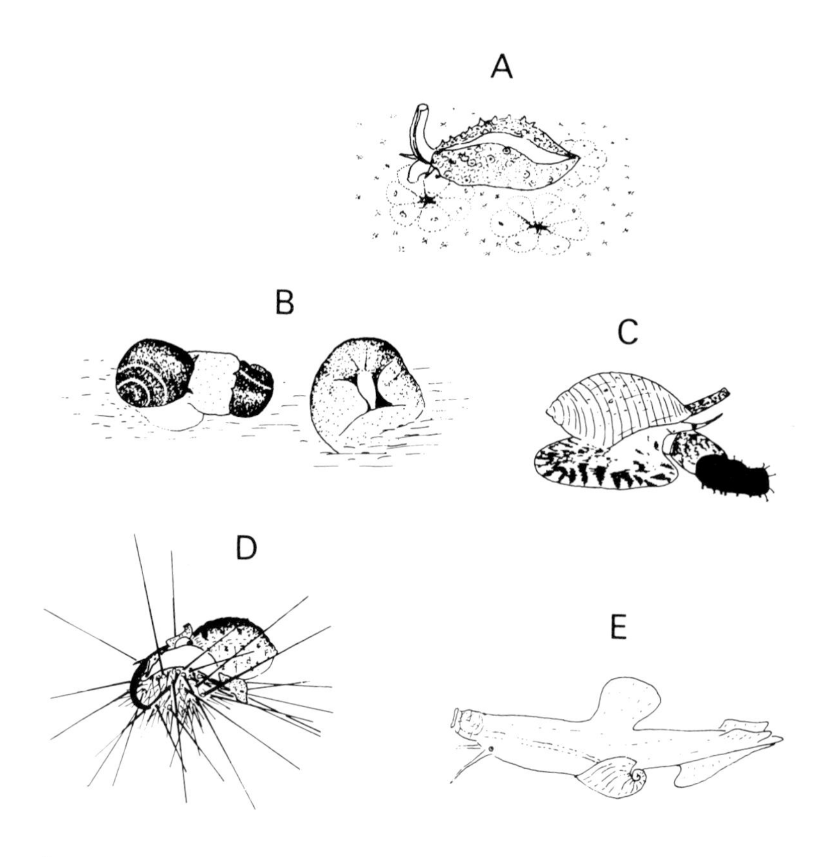

from a venom gland. Some terebrids lack venom, but rapidly envelop their enteropneust prey with the extended proboscis. These diverse methods of attacking prey are summarised in a scheme for the possible radiation of feeding behaviour in predatory prosobranchs (Figure 2.9).

Opisthobranchs are among the most highly evolved gastropods,

Figure 2.8: Carnivorous Neogastropods. A. *Buccinum undatum* (4-5 cm) wedging the lip of its shell between the valves of its prey *Cerastoderma edule*. After Nielsen (1975). B. *Urosalpinx cinerea* (1-2 cm) drilling an oyster. A pit under the front portion of the foot contains an eversible glandular pad, the accessory boring organ, whose acid secretion containing carbonic anhydrase etches the prey's shell (lower right figure). Periodically, the softened material is removed by the radula (upper right figure), laying bare more shell for etching. After Carriker (1969). C. *Conus purpurascens* (6 cm) harpooning and ingesting the goby *Elacatinys punctioulatus*. In the upper figure, the proboscis is being shot out towards the goby. In the lower figure, the proboscis has retracted and the narcotised goby is being swallowed. The mouth of *C. purpurascens* is immediately behind the head of the fish and the short eye-stalks can be seen near the rim. The buccal and anterior oesophageal regions are greatly distended by the bulk of the prey. After Kerstitch (1984)

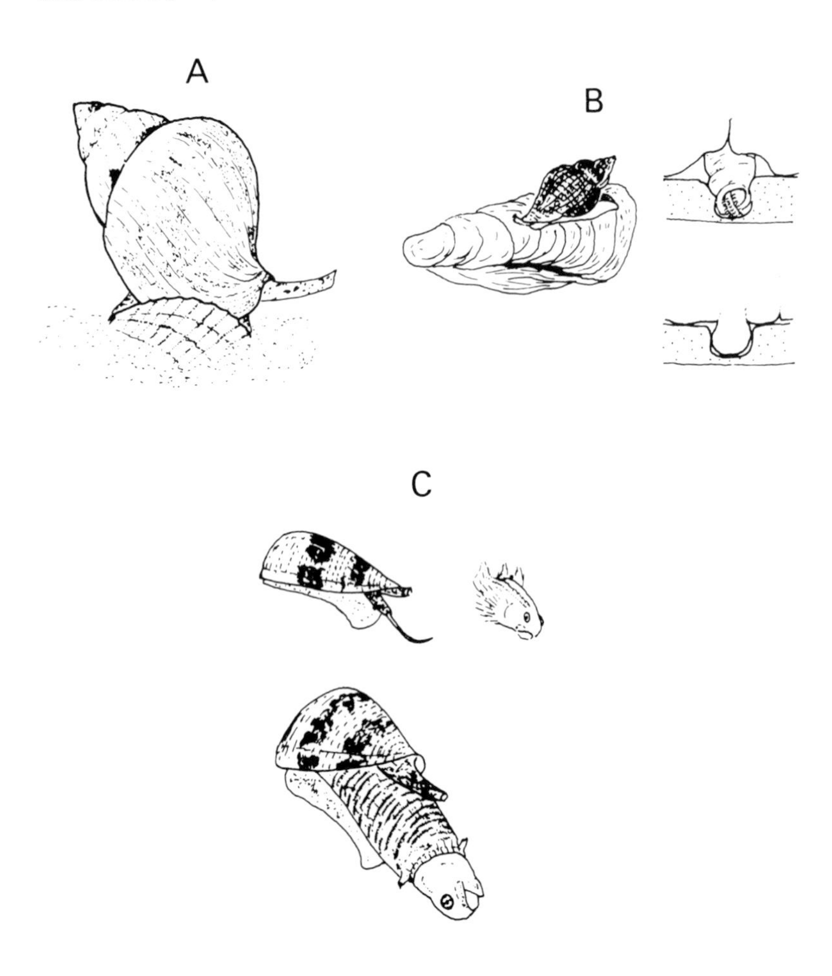

Figure 2.9: Possible Radiation of Feeding Methods in Predatory Prosobranch Gastropods. Predatory gastropods probably developed from ancestors feeding on sedentary animals such as sponges, coelenterates and ascidians. Radiation involved better methods of seeking and attacking mobile prey. Examples illustrated are from the families Eratoidae (proboscis probing), Nasariidae (proboscis probing and pedal manipulation), Naticidae (shell drilling), Buccinidae (shell wedging), Fasciolariidae (shell chipping), Olvidae and Harpidae (pedal asphyxiation), Terebridae (proboscis envelopment), Cymatiidae (toxic secretion), Conidae (toxic injection). Scheme after Taylor *et al.* (1980); examples are original, representing extant forms with appropriate feeding methods and do not follow any phylogenetic sequence. Drawings are not to scale

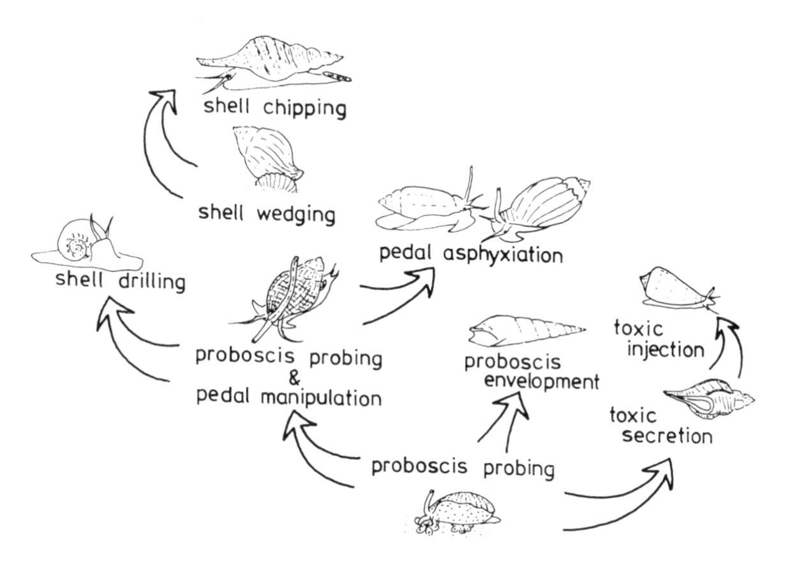

with both specialised herbivores (section 2.1.6) and predators. Among the latter, the order Nudibranchia is represented in shallow marine habitats throughout the world. Nudibranchs range from sponge-grazers (Figure 2.10A) with broad radulae to those with blade-like teeth for cutting into soft prey such as anemones, or with a buccal pump for sucking out the tissues of barnacles, bryozoans and tunicates. The planktonic order Gymnosomata is a group of highly specialised predators that feed specifically on filter-feeding thecosomes (section 2.1.4). *Clione limacina*, for example, feeds on *Spiratella helicina*, holding the prey's shell with tentacular lobes of the buccal lining while grasping and extracting the flesh with protrusible chitinous hooks (Figure 2.10B). Bullomorph opisthobranchs (Figure 3.2) are benthic, shelled gastropods with a large oesophageal gizzard lined with tooth-like plates. Bivalves, gastropods,

Figure 2.10: Predatory Opisthobranchs. A. Nudibranch *Chromodoris quadricolor* (4 cm) grazing a sponge. B. Gymnosome pteropod *Clione limacina* (1-2 cm) feeding on thecosome *Spiratella retroversa*; *C. limacina* lacks a shell and is transparent. Buccal tentacles hold the prey's shell while a pair of toothed hooks is inserted into the aperture to grasp and withdraw the flesh. After Lalli (1970)

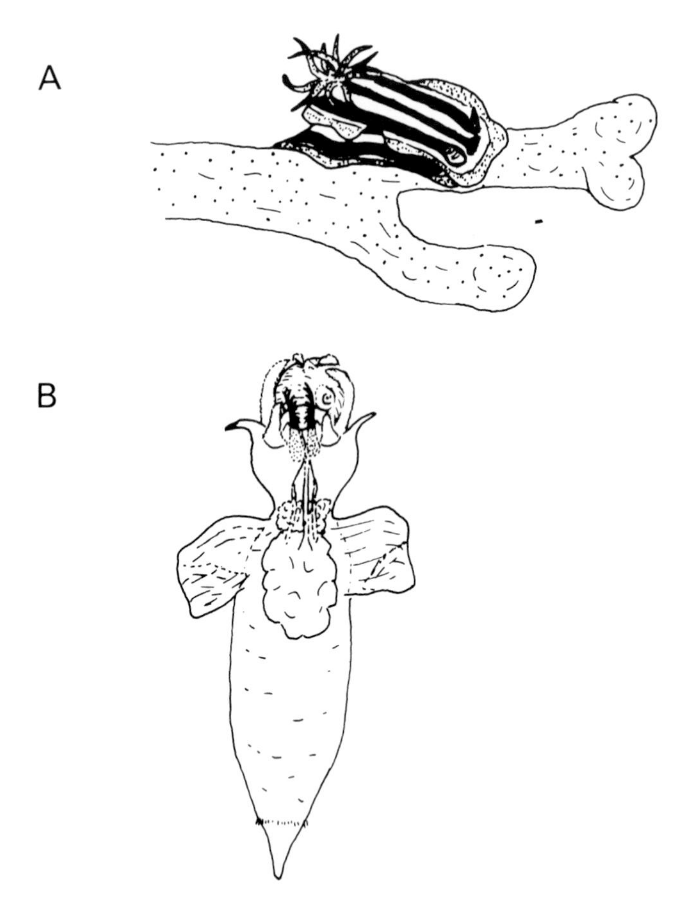

foraminiferans and other invertebrates caught in the broad foot are swallowed whole and broken up by the gizzard.

2.1.6 *Feeders on Cell Sap*

The opisthobranch order Sacoglossa (= Ascoglossa) contains species with one radular tooth per row. The tooth is used as a stylet for piercing algal cell walls (Figure 2.11). Some species have teeth shaped precisely to deal with particular algae, which are therefore the sole constituents of the diet. *Elysia viridis* mostly feeds on green algae such as *Cladophora.* The anteriormost (oldest) tooth on the radula (Plate 2.2A) functions as a serrated blade, cutting through the algal cell wall as the radula is protracted. This contrasts with most other gastropods, in which the teeth operate as the radula is retracted (section 1.3). Pressure acting on the anteriormost tooth as it cuts the algal cell wall will make it flex against the tooth immediately behind. Each tooth on the radula slots into a groove on the upper surface of the one in front, the whole series locking together as a rigid support for the tooth at the tip (Plate 2.2B). The terminal tooth is therefore analogous to the blade of a scalpel and the interlocking teeth to its handle. Green algae are a common food of sacoglossans, the large cells of these plants being lanced individually as the filament or thallus passes between the lips. The muscular pharynx pumps the sap out of the algal cell and into the stomach.

Figure 2.11: Sacoglossans and their Radular Teeth. One tooth is functional at a time and is used as a stylet for piercing algal cells. A. *Stiliger bellulus* (1 cm). The radular tooth is about 40 μm long and shaped like a scalpel blade. B. *Limapontia depressa* (2.3 cm). The radular tooth is about 70 μm long and is shaped like a gauge. After Thompson (1976)

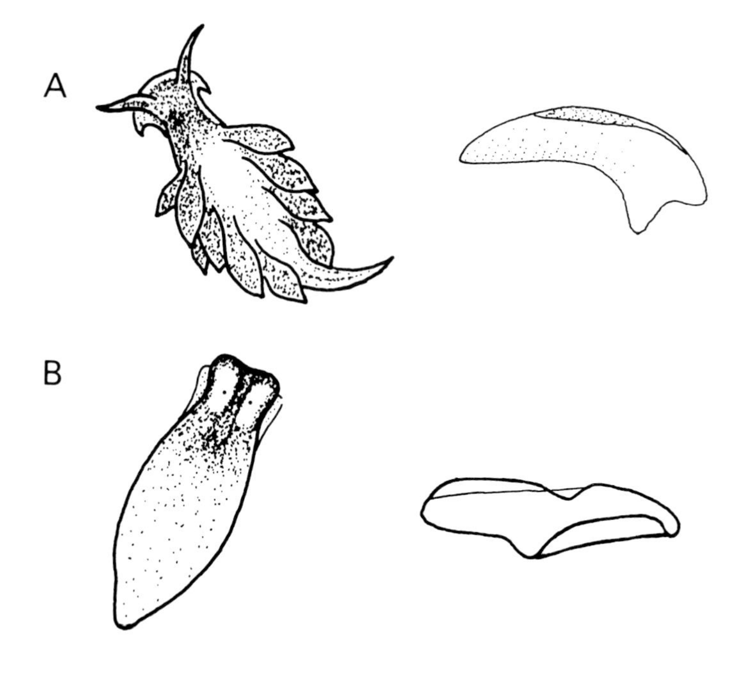

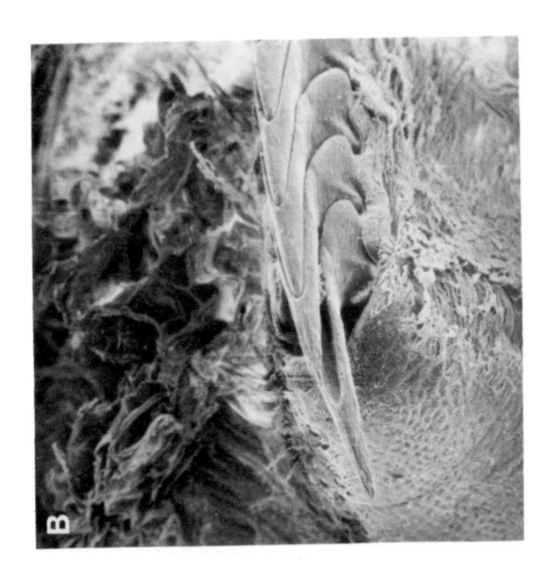

Plate 2.2: The Radula of *Elysia viridis*. A. Anteriormost tooth, showing serrations on the cutting blade and the (upper) groove into which slots the next (more posterior) tooth when pressure bears on the blade. B. The longitudinal series of interlocking teeth forms a rigid lever, transmitting force to the front tooth as it cuts through algal cell wall. Scanning electron micrographs by N.W. Runham. Magnification: A = × 3500, B = × 1000

2.1.7 Photosynthetic Symbiosis

Among sacoglossans, the Elysiidae are able to retain intact chloroplasts in the cells of the digestive gland, which ramifies into dorsal flaps (parapodia) (Figure 2.12). The chloroplasts remain functional for up to several weeks and net oxygen evolution has been observed in several elysiid species (Muscatine and Greene, 1973). An extensive network of dorsal veins, extending from the parapodia and coalescing at the pericardium, possibly functions as a 'negative gill' for the uptake of carbonate, used by the chloroplasts, and excretion of excess oxygen produced by photosynthesis (Stirts and Clark, 1980). Maintenance of symbiotic chloroplasts probably costs energy (Clark *et al.*, 1979), and the ratio of gain (production of photosynthates) to cost (maintenance metabolism) may be affected by the environmental temperature regime, accounting for the predominance of the symbiosis among tropical species (Stirts and Clark, 1980). Reputedly confined to the Elysiidae, chloroplast symbiosis may be a widespread, primitive characteristic of the Sacoglossa (Clark and Busacca, 1978).

Figure 2.12: *Elysia viridis* (2-3 cm). A. Ramification of the digestive gland tubules into the parapodia (dorsolateral Flaps). a=Anus, bc=buccal cavity, dg=digestive gland, pa=parapodium, s=stomach. B. Symbiotic chloroplasts occur within the cells of the digestive gland. cp=chloroplast, dg=digestive gland, sc=secretory cell, sv=secretory vacuole. After Muscatine and Greene (1973)

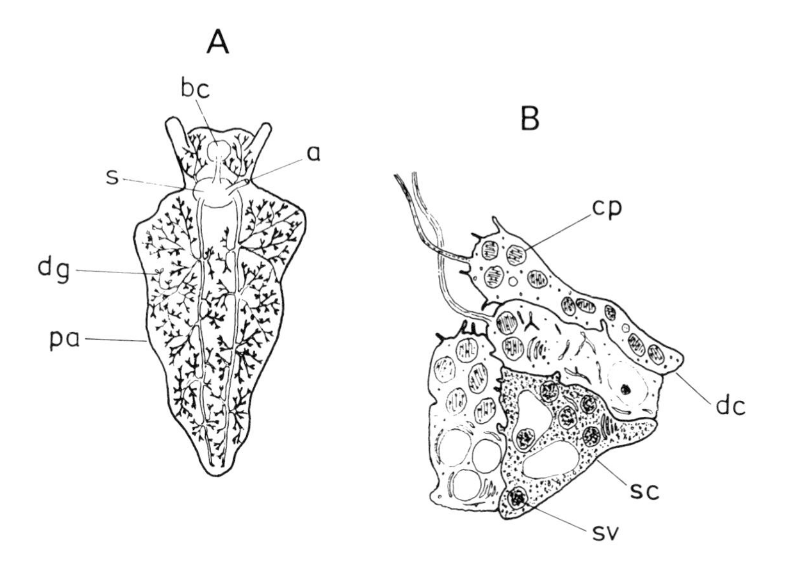

A few tropical aeolid nudibranchs, such as *Pteraeolidia ianthina*, contain symbiotic zooxanthellae (dinoflagellates) that translocate photosynthates to the host (Kempf, 1980). How the zooxanthellae are acquired is unknown as they do not occur in the prey and are not present in the larval stages of the host.

2.1.8 Parasites

Two families of small, shelled gastropods, the Eulimidae and Pyramidellidae, are mobile ectoparasites, the former on echinoderms and the latter on coelenterates, sedentary polychaetes, crustaceans, bivalves and gastropods. Both families were once placed in the mesogastropod superfamily Aglossa, so named because they lack a radula, but the Pyramidellidae bear closer resemblance to opisthobranchs (Fretter and Graham, 1949). Penetrating the body wall with digestive enzymes secreted from the proboscis epithelium (Eulimidae) or with stylet-shaped jaws (Pyramidellidae), these ectoparasites insert their long proboscis into their host's body cavity and suck the fluid (Figure 2.13A).

The Entoconchidae are endoparasitic prosobranchs, ranging from shelled forms such as *Stilifer* to totally modified forms unrecognisable as molluscs, such as *Enteroxenos*. Embedded within the arms of asteroids, *Stilifer* induces the host tissues to produce an enveloping gall. The long proboscis extends deeply into the host and ingests body fluids (Figure 2.13B). *Enteroxenos* parasitises holothurians, attaching to the gut when young and later living in the coelomic cavity as a white, worm-shaped animal up to 15 cm long and totally lacking in molluscan features (Figure 2.13C). The extensive body cavity serves as a uterus for the numerous eggs, and in the absence of a mouth and gut, food is absorbed across the body wall. *Enteroxenos* embryos pass through the veliger stage, complete with protoconch (Figure 2.13C), revealing their identity as gastropods.

2.2 Feeding Behaviour

2.2.1 Introduction

According to the energy maximisation premise (see Preface), feeding behaviour will tend to maximise the acquisition of food per unit metabolic cost. This, however, should be regarded only as an underlying principle; the need to restrict foraging or modify the

Figure 2.13: A. *Melanella alba* Parasitising the Starfish *Neopentadactyla mixta*. Upper figure shows the snail (2 cm) with its proboscis extended to the arm of its host. Lower figure illustrates the proboscidial mechanism of penetration. The proboscis unfolds from the proboscis sheath under the force of hydrostatic pressure (small arrows). Enzymes secreted from the outer surface of the proboscis soften the host's connective tissue. Thread-like retractor muscles enable the proboscis to withdraw when necessary. After Smith (1984). B. *Stilifer linckiae* (1-3 mm) parasitises the starfish *Linckia multifora*, sending a long proboscis into the host's water vascular system. The shell is surrounded by a sleeve-like reflexed extension of the proboscis epithelium and the whole parasite is enclosed within the host's tissue, which has been stimulated to form a gall. Females (4-5 mm shell height) are larger than nearby males, who copulate with them using a long penis. After Lützen (1972). C. Female *Enteroxenos oestergreni* (10 cm) attach themselves to the oesophagus of the holothurian *Stichopus tremulus* (upper figure). The body becomes a uterus for the numerous eggs (middle figure). Tiny, sac-like males are brooded within the females' bodies and fertilise the eggs. Evisceration by the host expels the females and their eggs hatch into veliger larvae (0.5 mm) that are liberated into the seawater (lower figure) and infect new hosts. After Lützen (1979)

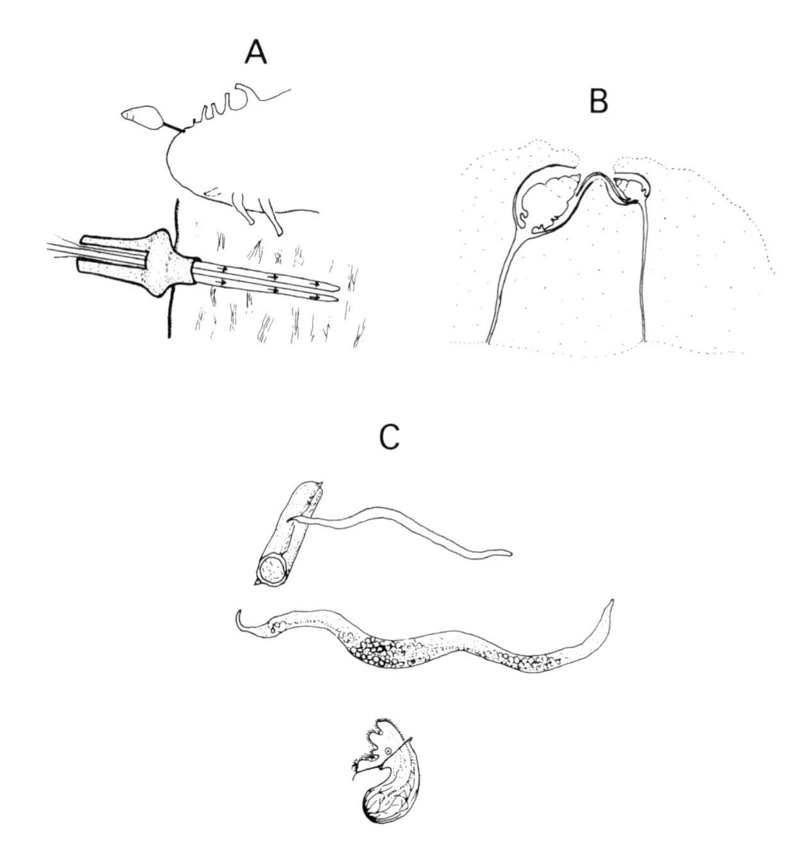

choice of food may arise in order to avoid excessive risks of death or interference from other animals.

2.2.2 Foraging Behaviour

Foraging behaviour, sometimes held to include all feeding processes from the search for food to its handling and digestion (Townsend and Hughes, 1981), is here regarded as those activities that enable the animal to encounter its food. Micro-algal grazers move slowly over the substratum, encountering food as they go; predators search for prey or wait for them to move within reach; suspension-feeders intercept water-borne food particles and deposit-feeders ingest sediment.

When gastropods move to encounter their food, two aspects of the movement may strongly influence foraging efficiency: the configuration of the trajectory and the speed. If food is depleted along the trajectory, at least for a time beyond the next foraging bout, the animal will forage more efficiently if it avoids recrossing its own path (Pyke, 1978). An animal frequently and randomly changing direction is very likely to recross its own path, but this likelihood is reduced if movement is in restricted directions. Non-random movement is therefore to be expected among grazing gastropods that simply crop food as they crawl over it. *Patella vulgata* grazes a meandering path but visits different parts of its home range on successive foraging bouts (Figure 2.14). *Patella longicosta* and *P. cochlear* on South African shores defend a 'garden' of red algae from other grazers. The algae are grazed systematically, leaving a reticulated pattern of algal turf. This grazing pattern probably maximises the sustainable yield from the algae, which are eradicated when other grazers intrude after experimental removal of the territory-holder (section 7.2.2).

Because prey usually occur in patches, a foraging predator should move in ways that reduce the time spent between patches and maximise the time spent within patches of food. Straighter, more 'directional' movement will take the animal quickly through areas devoid of food, whereas convoluted movement, even if random, will prevent the animal from progressing away from a patch of food. Despite recrossing its own path, the predator is more likely to encounter prey by meandering within the patch of food than by moving away. *Nucella lapillus* feeds on barnacles and mussels, which often occur in patches on the shore. Experimentally starved *Nucella* move in a highly 'directional' manner, following a

Figure 2.14: Return Paths from Grazing Site to Home Scar Taken by *Patella vulgata* on Different Days. The radius of travel ranges from about 10 to 30 cm. Different sectors of the home range are used on different days. Side-to-side sweeping of the head causes a zig-zag grazing pattern. Grazing ceases on the straighter, homeward sections of the path. After Newell (1979), based on Cook *el al.* (1969) and on Funke (1968)

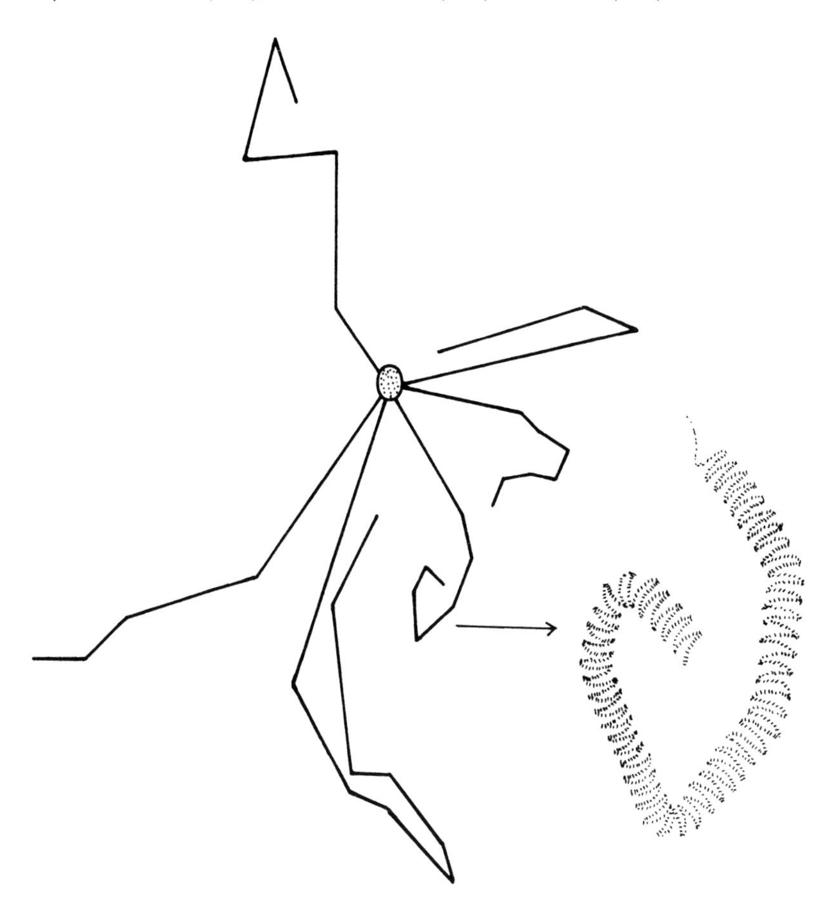

straighter path than recently fed snails, which crawl in a randomly tortuous trajectory (Figure 2.15).

Speed of movement is also critical; faster movement will increase the encounter rate with food but is energetically more expensive and the greater cost must be balanced by a sufficiently higher gain. Micro-algal grazers move over a continuous surface of food and, since the buccal apparatus (section 2.1.2) is mechanically limited in its functional capacity, fast crawling would waste

Figure 2.15: The Effect of Hunger on the Tortuosity of the Search Path of *Nucella lapillus*. The histograms represent the frequencies of angular deviations from the previous direction in successive 3 min intervals while searching. A. Snails recently fed, B. Snails starved for 90 days. Hungrier snails tended to move in straighter paths (little angular deviation between successive moves). After Hughes and Dunkin (1984a)

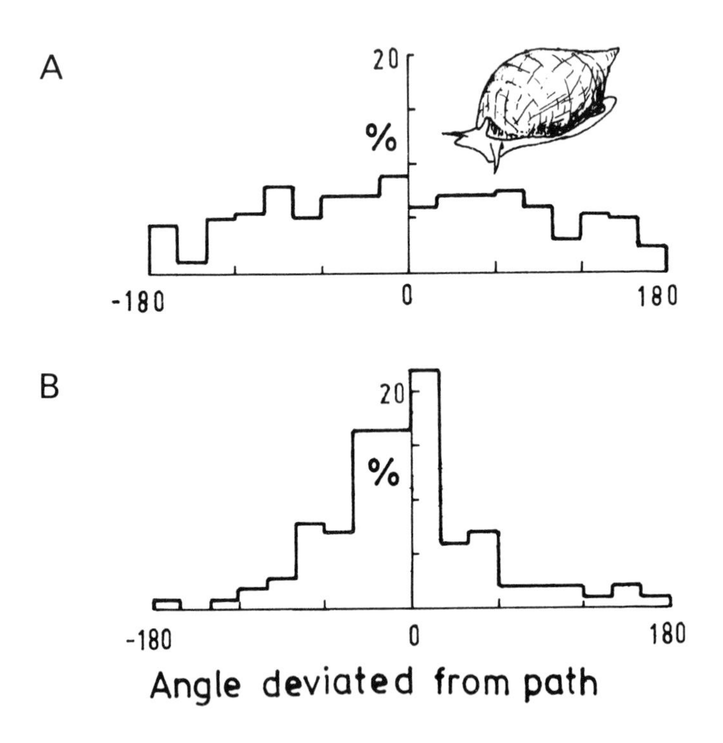

energy. *Nerita atramentosa* and *Bembicium nanum* have broad intertidal distributions on Australian shores, yet even though individuals living high on the shore have less time to feed during the tidal cycle than those lower down, they graze at the same rate, which, although slow, is probably the maximum physically possible (Underwood, 1984).

Adjustment of the grazing rate may, however, occur in other gastropods. Individuals of the New Zealand trochid *Melagraphia aethiops* living high on the shore crawl 40 per cent faster and scrape the radula 10 per cent faster than those at low shore levels. Moreover, high-shore individuals feed continuously when covered by the tide, whereas low-shore individuals feed intermittently (Figure 2.16). By increasing its ingestion rate and feeding continuously when conditions are suitable, *Melagraphia* compensates for the shorter time available for feeding high on the shore. Similarly, in certain populations of *Patella vulgata*, limpets zoned high on the shore forage throughout 76 per cent of the period of tidal immersion, leaving their scars (home bases) about 18 min after submersion and returning 16 min before emersion. Limpets zoned low on the shore leave their scars about 53 min after submersion and return about 114 min before emersion, so foraging for only 60 per cent of the period of tidal immersion (Cassidy and Evans, 1981).

Patella vulgata adjusts both the speed and directionality of its movement during a foraging bout. On the outward and return journeys to its scar, the limpet moves at about 6 mm min^{-1} in a relatively direct line, minimising the time spent travelling to and from the feeding area. On reaching the outer region of its home

Figure 2.16: The Effect of Shore Level on the Feeding Behaviour of *Melagraphia aethiops*. ......, Percentage of population active at extreme high water neap level during the period of tidal submersion; --------, percentage of population active at extreme low water neap level during tidal submersion. o=Rate of radular movement of snails at EHWN; •=rate of radular movement of snails at ELWN. After Zeldis and Boyden (1979)

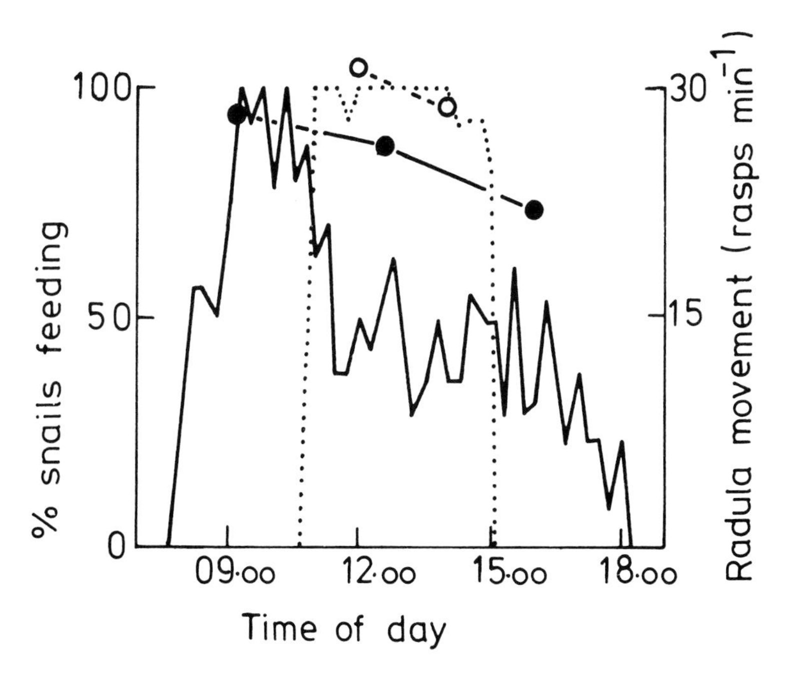

range, where algal food is more plentiful, *P. vulgata* slows down to about 0.8 mm min^{-1}, following a meandering path and swinging the head from side to side, so maximising the area of substratum grazed per unit of progression (Figure 2.14). Predatory gastropods tend to search slowly for prey; *Nucella lapillus* moves at a speed of about 0.5 m h^{-1} when foraging for mussels or barnacles (Hughes and Dunkin, 1984b), but acceleration over short distances may occur in response to stimuli from mobile prey. *Urosalpinx cinerea* doubles its crawling speed in response to the scent of the scallop *Argopecten maximus*, which can sometimes escape by flapping its valves and swimming away (Ordzie and Garofolo, 1980).

In times of food shortage, predatory gastropods tend to become quiescent, living off their energy reserves (section 4.3.5). The cassid *Galeodea echinophora* feeds on spatangoids in the fine offshore sediments of the Mediterranean Sea. Individuals have survived starvation for over 6 months buried beneath the sand in aquaria, and even in the presence of food they forage for only 2 to 4 h per day (Hughes, 1985a). Since spatangoids move slowly through the sediment, it may pay *Galeodea* to search in short bouts, saving energy and avoiding exposure to predators if prey are absent in the immediate vicinity. By the next foraging bout, prey might have moved into the area. Saving energy by capitalising on the movement of prey is taken to its extreme in ambush predators that do not search, but wait for prey to come within range of attack. For example, heteropods (Figure 2.7E) drift in the oceanic plankton, lunging at passing prey.

The feeding activities of others can be a reliable cue to a source of food. Dogwhelks are attracted by conspecifics engaged in the consumption of barnacles or mussels, probably sensing body fluids leaking from the wounded prey (section 2.2.3). A more sophisticated behavioural response has developed in the oyster-drill *Urosalpinx cinerea.* Starved *Urosalpinx* are repelled by the effluent of starved conspecifics but attracted to that of satiated ones (Pratt, 1976). This behaviour will increase foraging efficiency by directing snails towards aggregations of prey and away from unproductive areas.

Filter-feeding gastropods do not move in search of food, but the analogous behaviour is the pumping of water through the filtering device or the rhythmic deployment of mucous traps (section 2.1.4). Faster filtering uses more energy (Figure 2.17B), and animals may be expected to adjust their filtering rate to give the maximum yield of food per cost of filtering. Particles should not be accumulated faster than can be handled efficiently, and so the optimal filtering rate will decline as the concentration of suspended particles exceeds the threshold needed to saturate the feeding system. Appropriate adjustment of the filtering rate would cause it to rise as concentrations of suspended particles increase towards the saturation threshold and to decline beyond this, whereas ingestion rate would increase to an asymptote (Hughes, 1980). When fed on the haptophyte *Isochrysis galbana*, veligers of *Ilyanassa obsoleta* have filtering and ingestion rate curves with precisely these shapes (Figure 2.17A). When the larvae were fed on the green flagellate *Dunaliella tertiolecta*, their filtering rate declined monotonically, perhaps because the peak would occur below the lowest algal concentration used. The peaked ingestion rate curve of veligers fed on the diatom *Thalassiosira pseudonana* (Figure 2.17A) is unexpected and suggests that high concentrations of this alga interfere with the feeding process. Similarity between the shapes of observed filtering and ingestion rate curves and those predicted from the energy maximisation premise should, however, be interpreted with caution because mechanical saturation effects, without energy maximisation, could produce similar results. Deciding between these possibilities would require specific experimental design and this has not yet been done (Hughes, 1980).

Among gastropods using external mucus to trap suspended particles (section 2.1.4), the vermetid *Tripsycha tulipa* hauls its web more frequently when feeding in turbid than in clear water (Figure 2.17C). The energetic cost of hauling will be offset most when the web has become laden with food particles, and this will occur faster in turbid waters. Heavily laden webs take longer to ingest but this does not impede filtering because a new mucous web continues to be secreted during ingestion.

Application of the energy maximisation premise to deposit-feeders predicts that the ingestion of sediment should not become faster than is necessary to keep the gut fully packed, and that the speed of passage through the gut should allow the maximum extraction of nutrient per unit time. Ingestion rate is therefore expected to vary acording to the nutritional quality of the sediment, which will depend on the abundance of bacteria, blue-green algae, diatoms and other micro-organisms (section 2.2.3). Although some evidence corroborating the energy maximisation premise has been adduced from other invertebrates (Hughes,

Figure 2.17: A. Filtration Rates per Veliger Larva of *Ilyanassa obsoleta* Feeding on Increasing Concentrations of the Flagellate *Thalassiosira pseudonana* (•) or on the Diatom *Isochrysis galbana* (o). After Pechenik and Fisher (1979). B. Respiration Rate of *Crepidula fornicata* Increases Linearly as Filtering Rate Increases. Clearance rate, the volume of water cleared of algal cells per hour, is an index of filtering rate. The specimen was 124 mg in weight, the food was a diatom, *Phaeodactylum*, and the temperature was 15°C. After Newell and Kofoed (1977). C. Feeding Rate (Mean Intervals between Successive Hauls of the Mucous Web) of Three Individuals of the Vermetid *Tripsycha tulipa*. Mean intervals between hauls are shorter (feeding rate higher) when suspended matter is present in the seawater (solid bars) than when it has been experimentally removed (open bars). After Hughes (1985b)

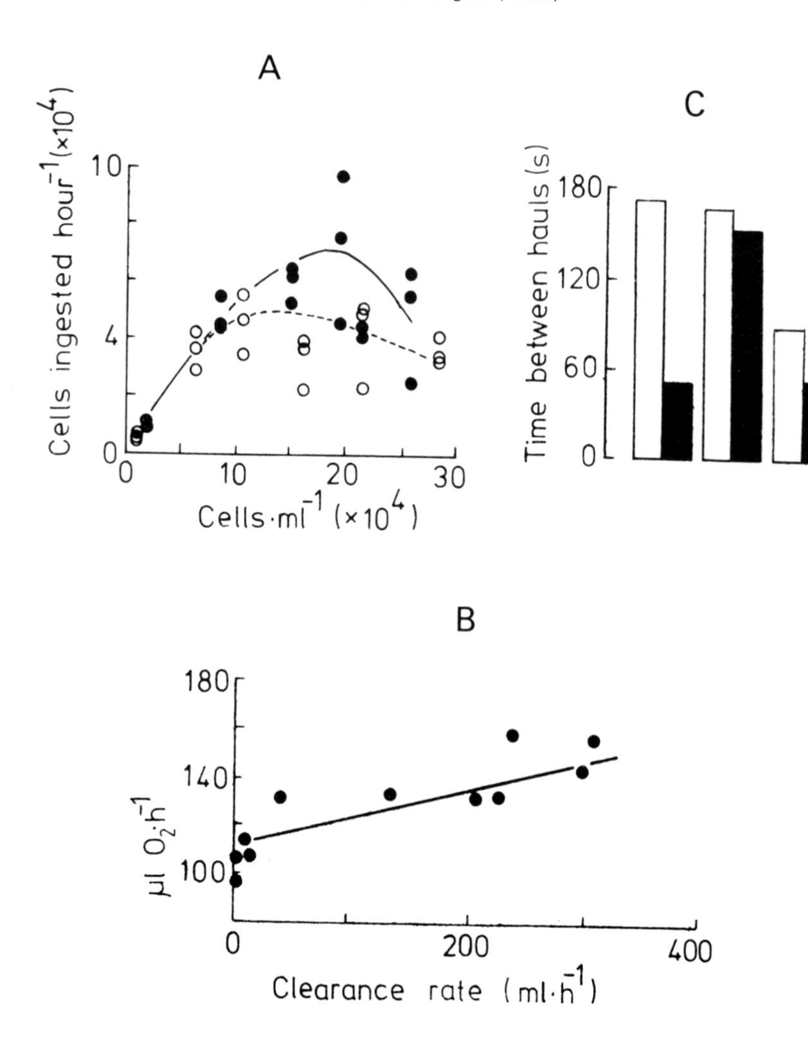

1980), appropriate experiments have yet to be done with gastropods.

When gastropods move about in search of food, they may be at greater risk to predation or environmental hazards than when hidden in a crevice, buried beneath the sand or clinging to a rock. Foraging bouts therefore tend to be restricted to the safest conditions. Cassids (Figure 2.7D) hunt for echinoid prey at night, hiding beneath the sand during daylight (Hughes and Hughes, 1980). *Nucella lapillus* searches for intertidal barnacles and mussels only during high tide when there is no risk of desiccation, but remains wedged within crevices in heavy seas to avoid being washed away (Menge, 1978). *Acanthina punctulata*, living on exposed Californian shores, stops searching for prey just before inundation by the rising tide, when waves pose the risk of dislodgement (Menge, 1974). Intertidal grazers cease feeding when the substratum begins to dry out during low tide, but show various responses to tidal immersion. On shores exposed to strong wave action, *Patella vulgata* grazes throughout high tide but on sheltered shores it becomes inactive while immersed, probably in response to predation pressure by the shore-crab *Carcinus maenas*. Similarly, on the Pacific coast of Panama, the limpet *Siphonaria gigas* clamps to its scar on the rock during high tide, and experimental eradication of the scar increases mortality, probably caused by puffer-fish (Garrity and Levings, 1981). The low-shore Californian limpet *Collisella limatula* feeds only while awash by the tide during daylight and only during low tide at night, so avoiding desiccation during the heat of the sun and predation by nocturnally hunting octopus (Wells, 1980).

2.2.3 *Diet Selection*

Probably all animals feed selectively. Grazing gastropods appear indiscriminately to ingest scrapings off the substratum, but the radula may remove chain-forming diatoms more easily than other forms (Nicotri, 1977) and *Lottorina tittorea* selectively remove certain algal sporelings (Watson and Norton, 1985). Deposit-feeders such as *Hydrobia* ingest sedimentary particles unselectively within the range of sizes that can be handled efficiently, but the size of the buccal apparatus restricts this range (section 7.2.3). The tropical dogwhelk *Thais carinifera* preys more frequently on the bivalve *Pelecyora trigona* than on *Anadara granosa*, whose valves are harder to force apart (Broom, 1983). Mechanical features of the animal and its food, therefore, will always impose some degree of differential feeding.

Food selection may also involve a behavioural response in which items are accepted or rejected on the basis of sensory information. *Littorina littorea* grazes preferentially on green rather than brown algae (Lubchenco, 1978); *Nucella lapillus* prefers barnacles to mussels (Hughes and Dunkin, 1984b); and even animals with monospecific diets are likely to prefer some items of the food to others, perhaps differing in nutritional status or ease of capture. According to the energy maximisation premise (see Preface), animals will select diets that maximise the energy assimilated per unit of metabolic cost, including all costs of handling the food from its pursuit, capture and ingestion to digestion (Townsend and Hughes, 1981).

The evolutionary perfection of the feeding apparatus associated with particular foods will account for mechanical, non-behavioural aspects of selective feeding and may be taken as basic fact. Behavioural selection of the diet in the manner predicted by the energy maximisation premise is more questionable, especially since immediate events such as risks of mortality may temporarily override the importance of energy maximisation. Nevertheless, accumulating evidence suggests that many gastropods do behave in ways that maximise the net gain of energy from their food.

The western North American dogwhelk *Nucella emarginata* grows fastest on an experimental diet of *Balanus glandula*, less well on *Mytilus edulis* and slowest on *Semibalanus cariosus*, and prefers these species in the same order when they are abundant on the shore (Palmer, 1983a). Similarly, *Natica maculosa* grows better on a diet of *Pelecyora trigona* than of *Anadara granosa*, and, when foraging on the intertidal mudflats of West Malaysia, preferentially feeds on *P. trigona*, using sensory information to discriminate among prey (Broom, 1983). *Aplysia punctata* shows a clear order of preference among the algae it normally encounters, and grows faster on diets of the more preferred species (Carefoot, 1967a). However, when presented with *Enteromorpha intestinalis*, an alga not normally encountered in its natural habitat, *A. punctata* prefers this species even to *Plocamium coccineum*, the most preferred naturally encountered alga, but a diet of *Enteromorpha* sustains poorer growth than one of *Plocamium* (sections 3.2 and 5.4.2).

A possible basis for selective feeding is the 'profitability' of the food. If handling activities (pursuit, attack, ingestion, digestion) for different foods have similar metabolic costs per unit time, then the profitability is the ratio of the energy assimilated from a food item to the handling time. Such will be the case when a predator chooses among different-sized individuals of a species of prey. The yield of energy will increase proportionately to the volume of the prey up to the point of satiation, but the handling time will also increase if larger prey are harder to catch or more resistant to attack. The precise relationship between yield, handling time and size will determine whether the smallest, the largest or some intermediate size of prey is the most profitable.

For *Nucella lapillus* feeding on *Mytilus edulis,* profitability increases with prey size (Figure 2.18A) and so the dogwhelk would be expected to prefer the largest mussels available. *Nucella lapillus* inspects individual mussels by crawling over them for more than an hour before either rejecting or drilling them. Contrary to expectation, *N. lapillus* prefers mussels smaller than the largest available (Figure 2.18A). The discrepancy would be accounted for if dogwhelks chose mussels that on average yield the most energy per unit handling time. In estimating the average profitability of mussels, account must be taken of the proportion of attacks in which a dogwhelk is disturbed or usurped by competitors attracted by olfactory stimuli from the damaged prey (section 2.2.2). This probability increases as handling time lengthens, so larger prey are devalued more by competition than smaller ones. By choosing mussels of intermediate sizes, dogwhelks may be selecting prey with the maximum average profitability in the face of intraspecific competition. Competitive devaluation of larger prey was invoked by Emlen (1966) to explain the preference of *Nucella emarginata* for *Chthamalus dalli* over the larger barnacle *Semibalanus cariosus. Chthamalus dalli* is small enough to be covered by the feeding dogwhelk, but large specimens of barnacles can accommodate several *N. emarginata* at once, and competing dogwhelks could devalue the profitability of the largest *S. cariosus* by as much as 30 per cent.

Boggs *et al.*, (1984) demonstrated the increased foraging efficiency gained when feeding on more profitable prey by supplying *Polinices duplicatus* with the clam *Mercenaria mercenaria,* whose valves had been ground down to half their natural thickness. The snails drilled more quickly through the thinner experimental shells, and the time saved was used to forage, so more food was ingested in the long term. Given a choice of experimental and natural clams, however, *P. duplicatus* could not identify the more profitable prey, having the ability to respond only to olfactory rather than tactile cues from the prey.

A highly selective diet will maximise the net rate of energy intake when the most profitable prey are abundant, but because of the extra time required to find the best prey, the predator would gain by becoming less selective when these are rare. Palmer (1984) demonstrated this principle by caging *Nucella emarginata* on the shore with different abundances of large *Balanus glandula* (more profitable), small *B. glandula* and *Mytilus edulis* (less profitable). After a month the dogwhelks were released onto a rock populated by these prey and, with the aid of scuba equipment, were observed continually throughout the tidal cycle. As expected, dogwhelks conditioned to high densities of the most profitable prey preferentially attacked the large *B. glandula*, whereas those conditioned to low densities of prey attacked the different types in the proportions encountered. Further observations suggested that *N. emarginata* requires 7-10 days to adjust its 'expectation' of prey abundance in a new situation.

On gaining experience with a particular kind of prey, the predator may become more efficient in its pursuit, capture or ingestion and as a result the yield may increase or the handling time may shorten, causing the profitability of the prey to rise. Conversely, these skills will attenuate when the prey is rarely eaten, causing its profitability to fall. *Nucella lapillus* maintained on *Semibalanus balanoides* and subsequently presented with small *Mytilus edulis* drills in random positions through the shell of the first mussel attacked, but thereafter increasingly selects the thinnest part of the shell, so shortening the handling time. By the sixth or seventh mussel attacked, the dogwhelks have reduced the average handling time by some 27 per cent, increasing the profitability of small mussels by about 17 per cent.

When maintained on mussels and then presented with barnacles, *N. lapillus* usually penetrates the parietal plates of the first prey attacked, but develops an increasing tendency to enter between the opercular plates of subsequent prey, shortening handling time and doubling the profitability of larger barnacles (Hughes and Dunkin, 1984a). The profitabilities of small *M. edulis* and large *S. balanoides* therefore depend on the rates of attack and hence on their abundance. Because the profitabilities of large barnacles and small mussels may be transposed in rank (Figure 2.18B), it is possible that *N. lapillus* would 'switch' from a specialised diet of one prey to that of the other if the abundances of barnacles and mussels changed appropriately.

Dietary 'switching' has not yet been recorded in *Nucella lapillus*, but one of the first experimental demonstrations of frequency-dependent predation consistent with switching was made with the Californian muricacean *Acanthina spirata* (Murdoch, 1969). Prolonged feeding either on *Balanus glandula* or on comparably sized *Mytilus edulis* induced *A. spirata* to develop a preference for one prey, and, when subsequently given mixtures of both, the snails fed disproportionately on the familiar type (Figure 2.18C). These simultaneous data for different batches of *A. spirata*, however, give a static picture of frequency-dependent predation, and the sequential process of switching by individual snails remains to be demonstrated.

Of course, switching is only to be expected where alternative prey are of comparable profitabilities. *Acanthina spirata* strongly preferred *Mytilus edulis* to *M. californianus*, which is stronger and harder to penetrate, and maintained this preference whatever the maintenance diet or the relative abundances of the two prey offered (Murdoch, 1969). Examples of switching are of considerable interest because they are mechanisms with a potentially stabilising influence on predator–prey systems or with the capacity to maintain genetic polymorphisms by 'apostatic' selection.

Selection of prey by *Nucella emarginata* was shown to depend on some 'memory' of the previous abundance of prey and not on dietary conditioning (Palmer, 1984). *Nucella lapillus*, however, increases its preference during prolonged exposure to a particular prey, involving a conditioning process additional to the more immediate effects of learning discussed above. 'Ingestive conditioning' was first described for *Urosalpinx cinerea* feeding on barnacles and oysters (Wood, 1968) but had also been shown in an earlier study in which *Aplysia punctata* maintained for 80 days on *Enteromorpha* increased its preference for this species over several other algae presented (Carefoot, 1967a). Ingestive conditioning has since been demonstrated in various taxa and is congruent with the development of a 'searching image'. For example, *Aeolidia papillosa* feeds on a variety of sea anemones, but when maintained on *Sagartia troglodytes* it strongly prefers this species among others presented (Figure 2.19). The preference can be changed to *Actinia equina* when this prey is used for the maintenance diet

Figure 2.18: A. The Curve Represents the Profitability (Yield of Flesh per Unit Handling Time) of Mussels to *Nucella lapillus*. The histogram represents the sizes of mussels eaten by *N. lapillus* when given a choice of all sizes. After Hughes and Dunkin (1984a). B. Change in the Profitability of Barnacles (Open Bars) and Mussels (Solid Bars) to *N. lapillus* Caused by Dietary Training. LB: SM =large barnacles and small mussels; SB : LM=small barnacles and large mussels; E=experienced with the prey; N=naive to the prey. When profitabilities are comparable (LB : SM) training transposes the profitabilities of barnacles and mussels, but does not do so when profitabilities are dissimilar (SB : LM). From Hughes and Dunkin (1984b). C. Switching in *Acanthina spirata*. Batches of *A. spirata* were either trained to mussels, not trained, or trained to barnacles and subsequently given mixtures of the two prey with 17.7% barnacles, 50% barnacles and 83.3% barnacles, respectively. The percentage of barnacles eaten is plotted against the percentage offered. Deviations from the diagonal indicate that *A. spirata* trained to mussels chose disproportionately few barnacles and those trained to barnacles chose disproportionately more barnacles. After Murdoch (1969)

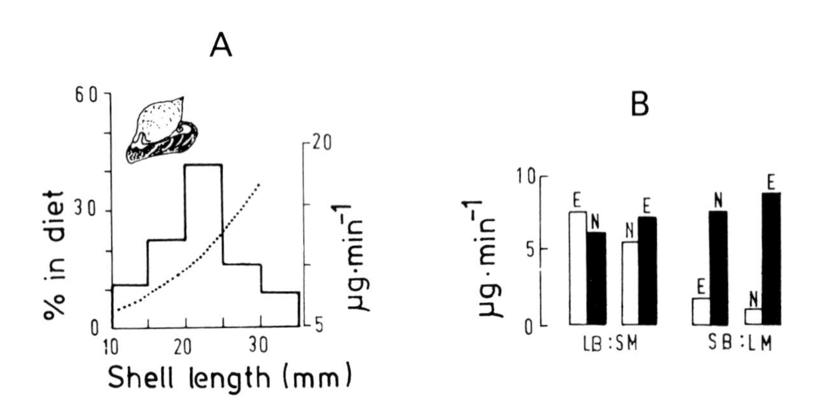

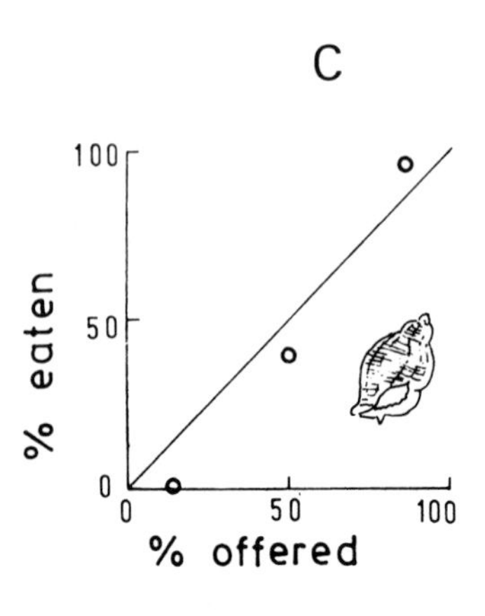

(Hall *et al.*, 1982).

Ingestive conditioning could delay the response of predators to changes in the abundance of prey. Some lag in this response will cause disproportionate dietary changes approximating to the idealised case of switching, but a more severe lag may cause the local extinction of a prey before preference for it has been lost. In the latter event, switching would have a destabilising influence on the predator-prey system, as seems likely with *Nucella lapillus* feeding on barnacles and mussels (Hughes and Dunkin, 1984b).

Although animal flesh does not vary in its nutritional quality as much as plant material with indigestible structural and defensive components, some organs are nevertheless easier to ingest or are richer in nutrients than others. When high-quality prey are abundant, selective ingestion will maximise the net rate of energy intake and moderately hungry predators may ingest only the most

Figure 2.19: Choice Chamber Used to Demonstrate Ingestive Conditioning in *Aeolidia papillosa*. Four types of anemone are placed in the chambers, one chamber remaining empty as a control. Water entering the back of the apparatus flows through the chambers and carries the scents of the anemones across the grid. A nudibranch starting at the bottom of the grid therefore has a choice of scent gradients to follow and takes about 10 min to move from the starting point to one of the chambers. Most individuals chose anemones on which they had previously been maintained, showing the effect of ingestive conditioning. After Hall *et al.* (1982)

profitable organs. Well-fed *Nucella lapillus* prefers the energy-rich digestive gland of mussels, abandoning tissues including the mantle rim and foot amounting to as much as 40 per cent of the total dry flesh weight of the prey (Hughes and Dunkin, 1984a). Similarly, *Polinices duplicatus* discards the siphons and mantle fringes of *Mya arenaria* (Kitchell *et al.*, 1981) and *Thais haemastoma* leaves the non-muscular tissues of oysters (Gunter, 1979).

In some cases, predators may position their attack to gain the most direct access to the nutritionally richest organs of the prey. The Australian dogwhelk *Dicathais aegrota* drills the shell of the limpet *Patelloida alticosta* directly over the gonad and digestive gland (Black, 1978). Whereas *Nucella lapillus* tends to drill the thinnest part of the shell of small mussels (described above), it most frequently drills another area, directly over the digestive gland, when attacking larger mussels. The different behaviour when attacking larger mussels may speed up the ingestion of the richest tissues before competitors are able to interfere or steal the prey (Hughes and Dunkin, 1984a). It is always possible, however, that the handling behaviour is determined merely by the geometry of the predator and its prey, as appears to be the case with the South African snail *Natica tecta* feeding on *Choromytilus meridionalis* (Griffiths, 1981).

Diet selection, like searching behaviour, may be constrained by other factors that prevent energy maximisation from being realised. Certain animals may need to diversify their diet to maintain an adequate balance of nutrients. This is especially likely among herbivores since plant material is more variable in nutritional quality than animal flesh (section 3.2). Probably as a result of nutrient balance, *Tegula* spp. from Californian shallow subtidal habitats maintain higher somatic and gonadal production when fed on a mixture of brown and red algae than on either alone (Watanabe, 1984). Nevertheless, these snails strongly prefer the giant kelp *Macrocystis* to all other algae, perhaps because this plant provides an adequate, although suboptimal diet and also affords protection from benthic predators. Other algae grow close to the sea-bed, and while feeding on them *Tegula* are heavily preyed upon by starfish.

The energy maximisation premise predicts that suspension-and deposit-feeders should select particles for ingestion only if the reward exceeds the cost of rejecting unprofitable particles (Hughes, 1980). Both the handling costs and the digestibility of particles are difficult to measure, and no critical data are available for marine gastropods. However, in the case of deposit-feeders which digest the bacteria and algal cells living among sedimentary particles, it is unlikely that particle selection, other than for size, would be energetically worth while. Instead, large amounts of sediment are processed rapidly, extracting the most accessible nutrients. *Hydrobia totteni* non-selectively ingests silt-clay particles within a certain size range (Levington, 1982) and extracts about 40 per cent of the bacterial and algal contents as the sediment passes through the gut (Lopez and Cheng, 1983).

2.3 Ingestion

2.3.1 Effect of Body Size

Metabolic rate is approximately proportional to body weight (section 4.2.1), and food requirement might be expected to follow suit. Ingestion rate, however, is largely determined by mechanical properties of the feeding apparatus and alimentary system, particularly by the cross-sectional area of lumina and by the area of filtering or absorptive surfaces. Surface and cross-sectional areas of a body are proportional to the square of any of its linear dimensions, and so are also proportional to its volume raised to the power $^2/_3$. If, therefore, proportions of the body remain constant (isometric) throughout growth, ingestion rate would be expected to be proportional to (body weight)$^{0.67}$.

Such a relationship holds for the freshwater limpet *Ancylus fluviatilis* and the freshwater snail *Planorbis contortus* (Calow, 1975). Wright and Hartnoll (1981), however, found that the ingestion rate of *Patella vulgata* grazing micro-algae is almost isometrically related to the body weight (exponent 0.9-1.0). Direct estimates of the ingestion rate of grazing gastropods are difficult to obtain, but the rate of faecal production is more easily measured and can be used as an indication of relative ingestion rate, assuming that, over sufficient time, input to the alimentary canal must be directly proportional to output. The exponent relating the rate of faecal production to body weight is recorded as 0.56-0.61 for *Nerita* spp. (Hughes, 1971a), 0.73 for *Fissurella barbadensis* (Hughes, 1971b) and about 0.79 for several species of *Patella.* Direct measurements of ingestion rate are made more easily with macrophagous species. The exponent relating daily ingestion rate to body weight has been estimated as 0.49-0.56 for *Nucella lapillus* feeding on *Mytilus edulis* (Bayne and Scullard, 1978b), 0.81 for *Polinices duplicatus* feeding on *Mya arenaria*, and 0.37 for *Clione limacina* feeding on *Spiratella retroversa* (Figure 2.20A,B). *Polinices, Clione* and *Patella vulgata* (see below) deviate considerably from the 'surface law', but the reasons for this have not been investigated; physiological and geometrical changes of the body could both be responsible.

2.3.2 *Effect of Food Quality*

Because of the greater quantities of indigestible material in plant

Figure 2.20: A. Ingestion Rate of *Clione limacina* as a Function of Body Size when Feeding on *Spiratella retroversa*. Both axes are logarithmic and the slope of the line is 0.37. B. Ingestion Rate of *Clione limacina* of 0.1-0.4 mg Dry Weight as a Function of Temperature when Feeding on *Spiratella retroversa*. The vertical scale is logarithmic. Between 3 and 17°C the Q_{10} is 3.0. Temperatures above 20°C are detrimental. C. Weight-specific ingestion rate of *Nucella lapillus* as a function of body size when feeding on *Mytilus edulis* at 15°C in March. Both axes are logarithmic and the slope of the line is −0.51. A and B after Connover and Lalli (1972), C after Bayne and Scullard (1978a)

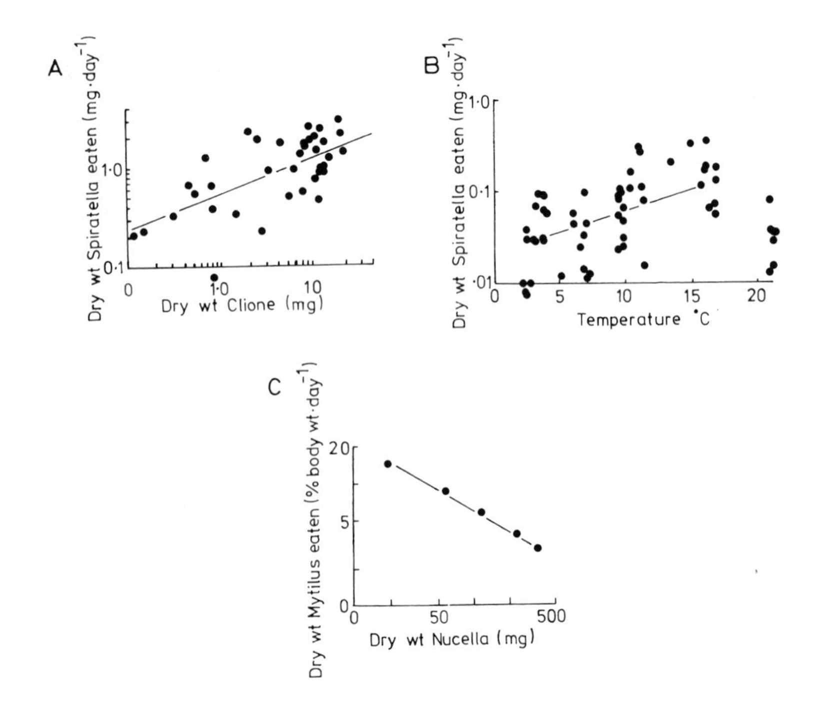

tissues, herbivores might be expected to process more food per unit time than carnivores. *Aplysia punctata* consumes its preferred food *Plocamium coccineum* equivalent to approximately 15 per cent of its dry body weight per day (Carefoot, 1967a), but micro-algal grazers have weight-specific ingestion rates similar to carnivores (Table 2.1). Comparison on a weight-specific basis, however, will tend to obscure interspecific trends because of the allometry between ingestion rate and body size (section 2.3.1).

2.3.3 Effect of Food Availability

Gastropods are sometimes short of food, in which case the average rate of ingestion will be less than maximal. Food shortage may cause an energy deficit and hence starvation, especially among carnivores exploiting unpredictable supplies of prey or among micro-algal grazers high on the shore where harsh environmental conditions curtail feeding or suppress algal growth. If starvation persists for too long, the animal may suffer physiological deterioration causing depressed feeding rates if food once again becomes available. But if starvation is for a shorter time, the animal may feed at an elevated rate in a homeostatic response to restore the energy balance. This has been shown for *Planorbis contortus* (Calow, 1975) and for *Nucella lapillus*, in which daily ingestion after 16 days' starvation was twice that after 4 days (Bayne and Scullard, 1978b).

Table 2.1: Ingestion Index for Gastropods, Defined as (Daily Food Intake)/(Body Weight) × 100. These figures do not take into account body size or seasonal changes of ingestion rate and are therefore approximate.

Species	Ingestion rate	Reference
Patella vulgata	1.3	Wright and Hartnoll (1981)
Fissurella barbadensis	12.2	Hughes (1971b)
Nerita tessellata	4.5	Hughes (1971a)
N. versicolor	4.0	Hughes (1971a)
N. peloronta	5.6	Hughes (1971a)
Littorina irrorata	3.5	Odum an Smalley (1959)
L. obtusata	2.3	Wright and Hartnoll (1981)
L. littorea	1.3	Grahame (1973)
Tegula funebralis	1.2	Paine (1971)
Pleuroploca gigantea	3.0	Paine (1963)
Polinices duplicatus	1.0	Edwards and Huebner (1977)
P. alderi	2.6	Ansell (1981)
Navanax inermis	6.2-9.5	Paine (1965)

Superabundance of food on the other hand causes satiation, but some gastropods continue ingesting food faster than the alimentary system can cope with at maximum efficiency. Even though the efficiency of digestion is lowered, the net gain may be higher if only the most accessible nutrients are extracted (section 3.2). When given an excess of *Enteromorpha*, the bullomorph *Haminoea zelandiae* ingests the food at such a rate that undigested algal filaments are voided with the normal faecal material (Rudman, 1971). Carnivores that do not swallow their prey whole respond to an abundance of food by ingesting tissues more selectively, ignoring the less nutritious or less accessible ones (section 2.2.3).

2.3.4 Effect of Temperature

Since active metabolism usually increases at higher temperatures within the normal physiological range of the animal (section 4.3.1), feeding activities are also likely to do so. A positive relationship between ingestion rate and environmental temperature has been recorded for *Polinices duplicatus* (Edwards and Huebner, 1977), *Clione limacina* (Figure 2.20B) and *Nucella lapillus*, which had a Q_{10} of 1.7 at 9-15°C and 2-3 at 15-20°C when ambient environmental temperature was about 9°C (Bayne and Scullard, 1978b). Seasonal acclimation (dampening in response) to ambient temperature has been found in the filtering rate of *Crepidula fornicata*, but only slight adjustments to the rate of radular scraping have been noticed in *Littorina littorea* and none in *Patella* spp., which therefore show a largely unmodified, positive response of feeding rate to changes in temperature (Newell and Branch, 1980).

2.4 Neuronal Basis of Feeding Behaviour

Neuronal control of feeding in certain opisthobranchs and terrestrial pulmonates has attracted the attentions of physiologists because in these gastropods the nervous system and buccal mass have several features conducive to experimentation. Relatively few neurons are present and their large cell-bodies are readily seen at the surfaces of the ganglia, where they can be penetrated by microelectrodes recording electrical activity, or injected with fluorescent dye revealing axonal ramifications. Even when partially isolated from the body, the buccal mass with associated ganglia retains the ability to perform rhythmic feeding movements, greatly facilitating experimentation. The buccal muscles are anatomically distinct and innervated directly by motoneurons whose cell-bodies are located in the buccal ganglia. Neuronal connections can therefore be identified by simultaneously recording electrical activity from the cell-bodies and the surfaces of the muscles.

Using these facilities, Bulloch and Dorsett (1979a,b) investigated the endogenous cyclical activity of the buccal mass of *Tritonia hombergi*, which feeds on the soft-coral *Alcyonium digitatum*. To attack its prey, *T. hombergi* slightly protrudes the buccal mass through the mouth, exposing the jaws and the radula. It then seizes the food with the jaws and undergoes a series of biting cycles, gradually cutting through the coral. Following each bite the radula makes two types of retraction movement, one carrying the food back into the buccal cavity and the other releasing it and pushing it into the oesophagus. The biting movements of the jaws eventually succeed in detaching a piece of the *A. digitatum* colony, which is then pulled in as the radula folds and moves back into the buccal cavity. The radula releases the food, and dorsal movements of the odontophore help to push it into the oesophagus.

Complete cycles of activity can be elicited from the surgically isolated buccal mass, having severed all nerves to the body wall. Cell-bodies in the buccal ganglia have been identified as motoneurons specifically associated with the protraction, retraction or flattening phases of buccal activity (Figure 2.21A). Motoneurons fire bursts of action potentials during appropriate phases of the buccal cycle (Figure 2.21B) and are directly responsible for the associated muscular activities. Synchronous firing within each group is assisted by electrical coupling of the motoneurons, and the phasic activity of all three groups is driven endogenously by three interneurons within the buccal and cerebral ganglia. A number of other neurons within the cerebral and buccal ganglia act as 'higher order' interneurons, which can initiate the whole feeding programme; however, the control of these cells is not understood.

Neurophysiological studies of feeding behaviour in other gastropods, including *Aplysia*, *Pleurobranchaea* and *Navanax*, are reviewed in Kandel (1976). Surgical lesion of ganglia and connectives has shown that orientation (appetitive) behaviour elicited by

Figure 2.21: A. Cycle of Buccal Activity in *Tritonia hombergii* when Feeding. Phases 1-3 may be iterated numerous times before returning to rest. The buccal mass is depicted with the oesophagus uppermost and the mouth region lowermost. IL = Inner lip, M=mouth region, OD=odontophore, OE=oesophagus, RA=radula. B. Cycles of Buccal Activity are Accompanied by the Integrated Activity of Specific Neurons in the Buccal Ganglia. Illustrated are electrical recordings from four neurons involved in the three phases of the buccal activity cycle. After Bulloch and Dorsett (1979a)

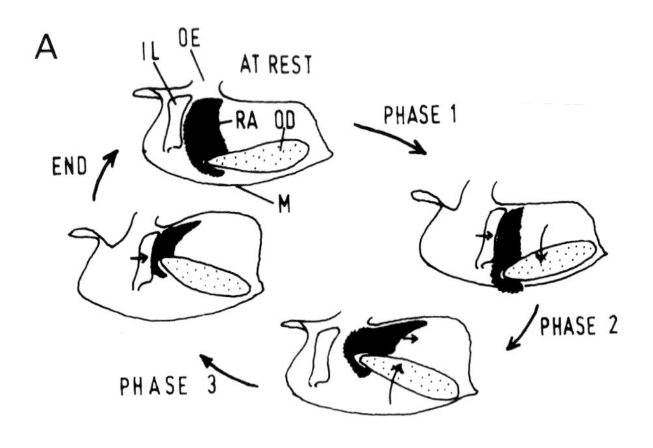

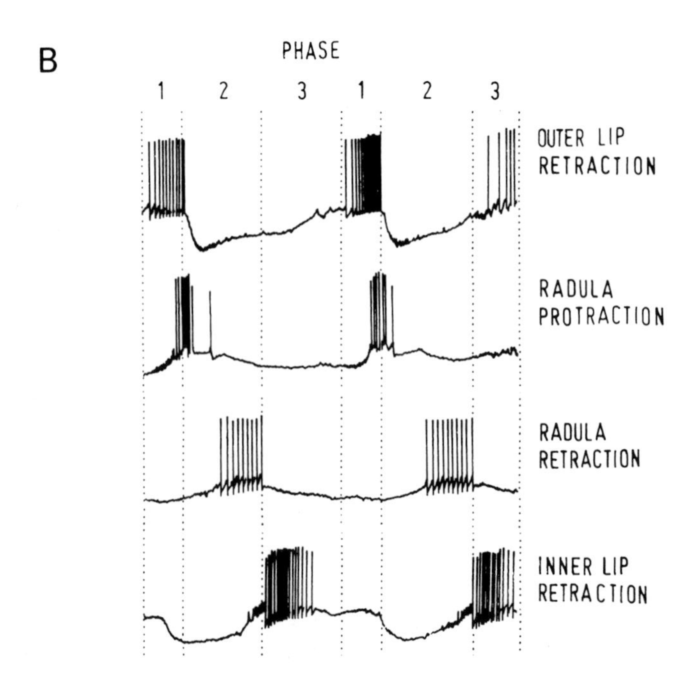

olfactory stimuli is under separate control from feeding (consumatory) behaviour. When pieces of the alga *Porphyra* are presented near the lips and posterior tentacles, *Aplysia californica* sways its head from side to side. This appetitive behaviour enables *A. californica* to locate the food and is controlled by neurons in the pleural and pedal ganglia (Kupferman, 1974). Consumatory biting, probably triggered by sensory information from the lips, transmitted via the cerebral ganglion and cerebrobuccal connectives, is controlled by neurons in the buccal ganglia.

Evidence suggests that at least some gastropods are capable of associative learning. *Pleurobranchaea californica*, a large carnivorous opisthobranch, learns to associate touching by a sterile glass rod (conditioned stimulus) with contact by homogenised prey tissue (unconditioned stimulus) after 20 trials at 30-60 s intervals (Mpitsos and Davis, 1973; Mpitsos *et al.*, 1978) and the learned behaviour is remembered for as long as 2 weeks. Avoidance behaviour is learned in fewer than 10 trials at 1-h intervals when presentation of the homogenate is accompanied by an electric shock.

3 DIGESTION AND ABSORPTION

3.1 Digestion

3.1.1 Alimentary Systems

Food is processed to various degrees among different gastropods as it passes from the buccal cavity, along the oesophagus, into the stomach. In micro-algal grazers, food is lubricated with mucus secreted by the oesophageal epithelium and by the salivary glands, which may also secrete amylases as in *Littorina littorea* (Fretter and Graham, 1962). Ciliated tracts carry the food-laden mucus to the stomach where carbohydrate-splitting enzymes, sometimes including cellulase, secreted by the digestive gland, partially break down the food. An area of ciliated ridges in the stomach wall, well developed in all microphagous gastropods, directs small particles to the openings of the digestive gland ducts. At least in some species, hydrostatic pressure generated by muscular contractions of the stomach forces the particles into these ducts. Muscular contraction of the ducts then moves the particles into the fine lumina of the digestive gland where they are phagocytosed and digested intracellularly. In other species, particles may be transported to the digestive gland by ciliary action. Larger, undigested particles are directed by the ciliated folds away from the digestive gland openings, towards the intestine. Waste material excreted by the digestive gland is transported by cilia outwards along the digestive gland ducts into the stomach, to join the rejected particles entering the intestine. Where the stomach merges with the intestine, rejected material and excreta are bound with mucus and rotated by ciliary action into a faecal rod. Passing along the intestine, the faecal rod is fragmented by muscular contraction of the intestinal wall and the pellets are expelled from the rectum. In deposit-feeding, filter-feeding and some grazing mesogastropods (e.g. Rissoacea, Cypraeidae and Strombidae) a rotating crystalline style (section 1.4) slowly liberates amylases and serves as a windlass, pulling food strings into the stomach. The globular protein nature of the style indicates that no proteases are present in the stomach of these gastropods (Yonge, 1930). Difficult to reconcile with this view, however, is the presence of a crystalline style in certain neogastropods such as *Ilyanassa obsoleta* (Ponder, 1973). The style is secreted and broken down on a daily cycle (Curtis and Hurd, 1981), but although *I. obsoleta* ingests large amounts of plant material and detritus that would be digested by amylases from the style, it also readily consumes flesh that requires proteases for digestion. Perhaps the style is present only when *I. obsoleta* is feeding on detritus.

A carnivorous diet is associated with more extensive extracellular digestion, especially the breakdown of proteins. Carnivorous prosobranchs have highly developed salivary and oesophageal glands (Figure 3.1) whose secretions lubricate and begin to digest the food on its way to the stomach. The predominance of extra cellular digestion is associated with simplification of the stomach, which often lacks any vestige of a ciliary sorting area.

Opisthobranchs show the greatest alimentay diversity (Thompson, 1976). Nudibranchs such as *Tritonia* feeding on soft, animal tissues have relatively simple alimentary systems in which the foregut and hindgut are merely conduits for the food and faeces. Salivary glands secrete a lubricating, enzyme-containing fluid that passes with the food into a simple sac-like stomach, which also receives enzymes from the digestive gland. Digested fluid is absorbed across the stomach wall and fine particles are taken up by the digestive gland. Gymnosomes also have a simple alimentary system, able to digest the flesh of their thecosome prey with remarkable efficiency (section 3.2). Flesh extracted from the prey's shell (Figure 2.10B) receives salivary secretions in the buccal cavity and passes directly along a simple, highly dilatable oesophagus to the digestive sac. This is a spacious chamber formed by the digestive gland and stomach, the latter being reduced to a small ciliated area (Lalli, 1970).

Other opisthobranchs have a more elaborate foregut equipped for temporary storage, mastication or digestion of food prior to its entry into the stomach. This has enabled bullomorphs to exploit large or hard-shelled prey or to become macro-herbivores. Among bullomorphs, the Aglajidae are active predators of polychaetes and of other gastropods, which are swallowed whole. In the case of gastropod prey, the shell passes intact through the gut. The oesophagus is dilated into a crop region (Figure 3.2C) where most of the digestion occurs. Muscular contraction of the stomach moves enzymes from the digestive gland to the crop. Conversely,

Figure 3.1: *Cassis tuberosa* (15-20 cm). A dissection to show the plan of the anterior alimentary system in which the visceral mass and part of the dorsal body wall have been removed. The highly extensible proboscis is used to drill and probe the echinoid prey. The proboscis glands secrete 0.1 N sulphuric acid, used to etch the prey's test. The salivary glands are of unknown function. The oesophageal gland is pleated, increasing its internal surface area, and is involved with digestion. BM=Buccal mass, OE=oesophagus, OG=oesophageal gland, OP=operculum, P=proboscis, PG=proboscis gland, RM=proboscis retractor muscles. After Hughes and Hughes (1981)

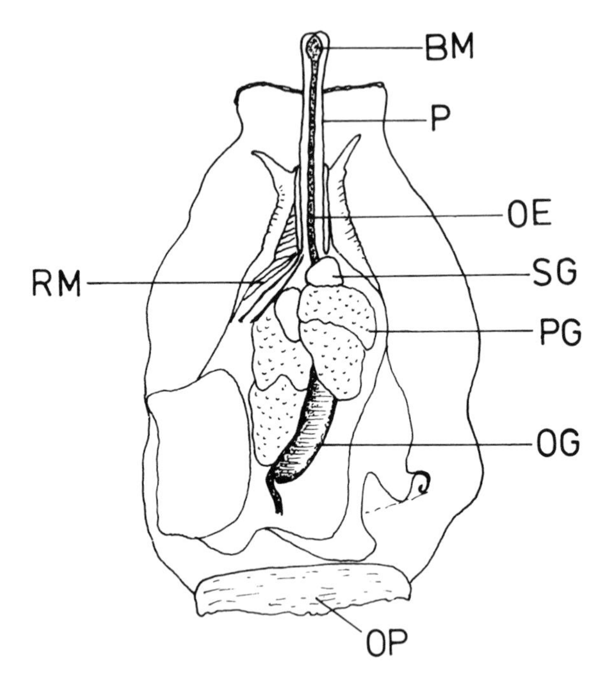

muscular contractions of the crop force digested fluid into the small, simplified stomach. Subsequent muscular contraction of the stomach moves the fluid into the digestive gland for absorption, and ciliary tracts carry enzymes and excreta out of the digestive gland into the stomach. The Philinidae are burrowing carnivorous bullomorphs that crush bivalves with a muscular gizzard lined with calcareous plates (Figure 3.2B). *Haminoea zelandiae* is a herbivorous bullomorph that commonly feeds on *Enteromorpha* (Figure 3.2A). The jaws grasp the algal filament and the radula snatches a short length and passes it to the crop region of the oesophagus, where it is packed ready for entry into the gizzard. Three spiny plates lining the gizzard regulate the influx of algal segments from

Figure 3.2: Bullomorph Alimentary Systems. A. *Haminoea zelandiae* (animal 2 cm). Large salivary glands secrete amylolytic enzymes that begin to digest the algal carbohydrate during passage through the oesophagus. The muscular gizzard contains toothed, horny plates that grind up algal filaments into a chyme that is passed to the stomach for further digestion. After Rudman (1971). B. *Philine auriformis* (animal 8 cm). Bivalve prey are swallowed by the muscular, bulbous buccal mass, and their shells are crushed by large calcareous plates lining the gizzard. The salivary glands are relatively small, as is typical of carnivorous gastropods. After Rudman (1972a).
C. *Melanochlamys cylindrica* (animal 2 cm). Polychaete and nemertine prey are sucked in by the bulbous buccal mass and stored in the crop, where they are digested by enzymes passed forward from the stomach. The soft-bodied prey do not need crushing and there is no gizzard. After Rudman (1972b). bm=Buccal mass, cr=crop, dd=digestive gland duct, gz=gizzard, i=intestine, oe=oesophagus, s=stomach, sf=salivary gland, pm=buccal mass protractor muscles

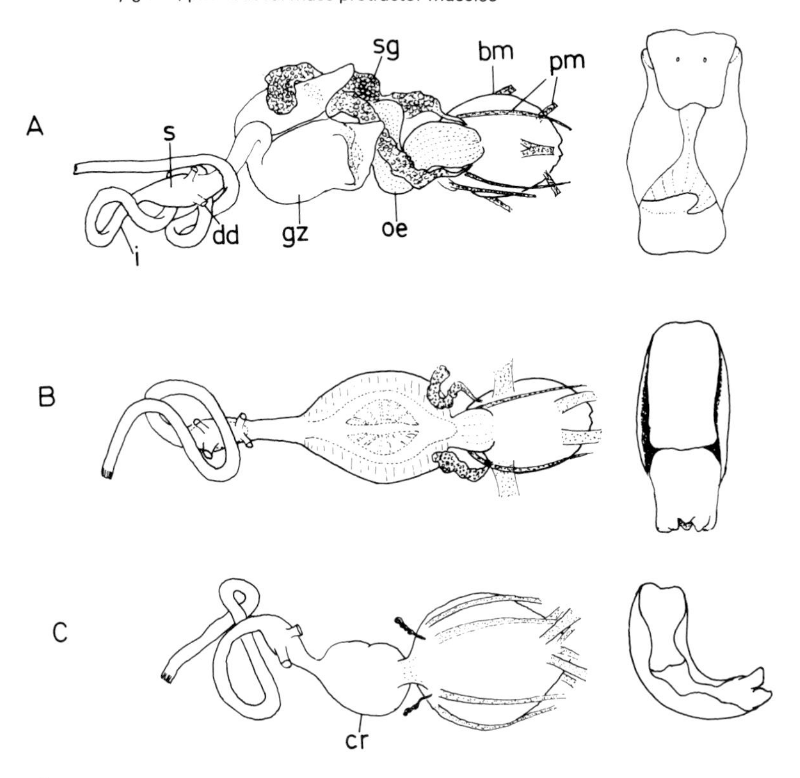

the crop and grind the food into small fragments. These are conveyed by muscular and ciliary action of the posterior foregut to the stomach.

Aplysiomorphs display the most elaborate modifications of the alimentary system for macro-herbivory (Figure 3.3A). Pieces of alga, cut by the jaws and radula, are stored in the crop before being processed by the double-chambered gizzard. The anterior chamber of the gizzard is lined with toothed, horny plates that triturate the algal fragments, breaking up the cell walls and facilitating digestion by enzymes that have been secreted by the digestive gland and passed forwards from the stomach. The posterior chamber is lined with setae that strain the pulp, allowing small particles and fluid to pass into the digestive part of the stomach but causing the rest to go directly to the intestine. The stomach receives ducts from the digestive gland ventrally and laterally, where a ciliary sorting mechanism assures that excessively large particles are directed, together with excreta from the digestive gland, away from the ducts and into the intestine. A channel bordered by flanges (typhlosoles) in the dorsal wall of the stomach carries undigestible algal fragments straight to the intestine. In the posterior wall of the stomach, this channel forms part of a caecum in which waste is compacted with mucus to form faecal rods.

Closely allied to the aplysiomorphs, the planktonic filter-feeding Thecosomata (section 2.1.4) have a similar gut (Figure 3.3B). The horny-plated gizzard crushes the cellulose walls of diatoms, but since only microscopic items are consumed there is no filter chamber. Both the aplysiomorphs and thecosomes lack cellulase, relying on trituration to release the digestible matter from algal cells.

3.1.2 Phasic Activity of the Digestive Gland

The digestive gland is usually far larger than the stomach, occupying much of the visceral mass, where it is recognisable as dark-brown tissue. Ciliated ducts connecting with the stomach branch into fine, blind-ending tubules around which are grouped digestive and secretory cells. Both are derived from columnar, basophilic cells near the ends of the digestive gland tubules in *Rissoa parva* (Wigham, 1976). Digestive cells are tall, cylindrical, with a basal nucleus and an 'apical brush' of microvilli, whereas the secretory cells are conical with a wide base and a narrow tip reaching to the lumen (Figure 3.4). Like the digestive cells, the secretory cells have a basal nucleus and apical microvilli, but they have denser cytoplasm with more extensive endoplasmic reticulum and Golgi systems, associated with protein synthesis.

Phasic activity of the digestive gland is probably widespread among molluscs and has been described in detail for *Littorina* spp. (Merdsoy and Farley, 1973; Boghen and Farley, 1974) and *Rissoa parva* (Wigham, 1976). Three phases of activity are recognisable.

Figure 3.3: A. *Aplysia* (10 cm). Algal food accumulates in the crop and is passed to the gizzard for trituration. Fine particles enter the digestive gland and larger particles and digestive waste are compacted into faecal rods in the caecum. After Thompson (1976). B. Thecosome Pteropod *Creseis acicula* (6 mm across the 'Wings', Posterior Third not Shown). The pteropod swims by beating the wings, which are covered in mucus that traps diatoms and other small particles. Ciliary fields transport the mucus to the mouth. Ingested diatoms are crushed by the gizzard and the fine, semi-digested particles are passed to the digestive gland. Digestive waste is compacted into faecal rods in the caecum. After Yonge (1926). A=Anus, BM=buccal mass, CE=caecum, CF=ciliary field, CR=crop, DG=digestive gland, GZ=gizzard, SG=salivary gland, SH=shell, W=wing

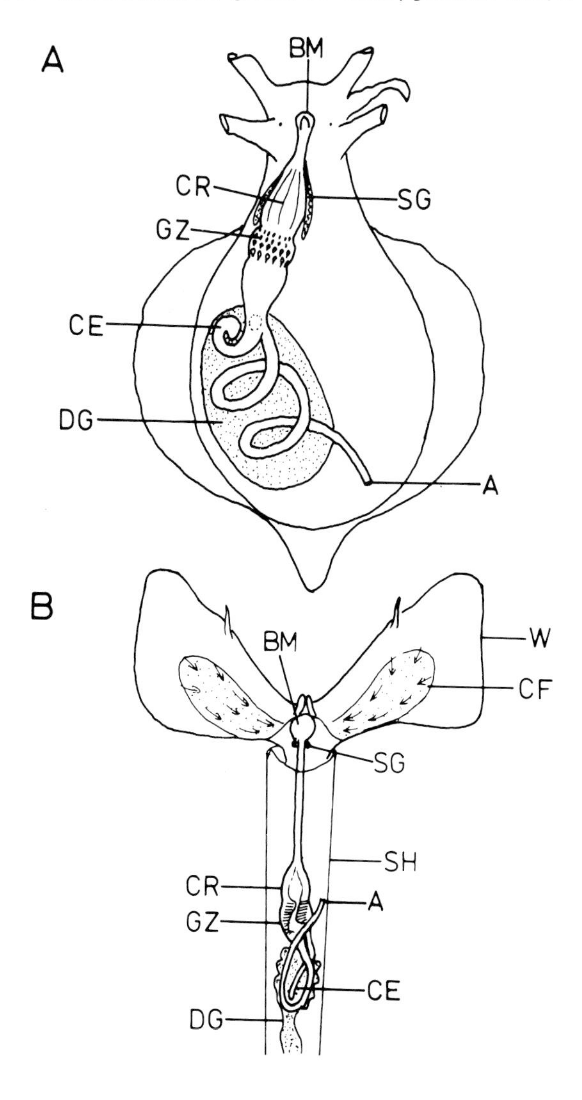

Figure 3.4: Cytology of the Digestive Gland, Based on *Littorina littorea*. Left, digestive cell in the absorptive phase; middle, secretory cell; right, digestive cell in the digestive phase. BM—Basal membrane, DV—digestive vacuole, ER—endoplasmic reticulum, EV—excretory vacuole, GI—Golgi apparatus, IV—intermediary vacuole, MC—mitochondrion, MV—microvilli, N—nucleus, PV—pinocytotic vesicle, SC—surface coat, SV—secretory vesicle. After Merdsoy and Farley (1973)

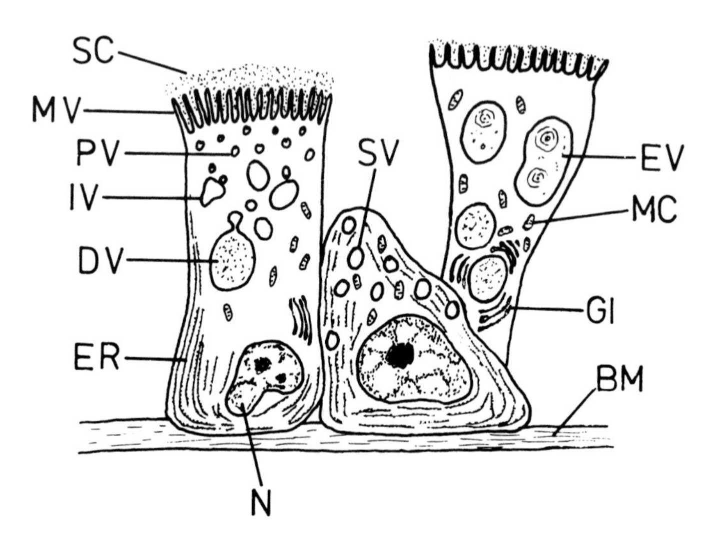

(1) Absorption: the digestive cells are tall and columnar, sometimes with a surface coat over the microvilli resembling that of absorptive cells in the mammalian intestine, but of unknown function; pinocytotic vesicles accumulate below the apical brush.
(2) Digestion: the digestive cells become shorter and more distended; the pinocytotic vesicles coalesce to form digestive vacuoles, first containing lipid droplets, and fibrillar and granular material but later becoming filled with homogeneous granules.
(3) Excretion: excretory vacuoles carrying the waste products of intracellular digestion burst into the lumen and the digestive cells regain their elongated shape.

The secretory cells also show phasic activity. During the absorptive to mid-digestive phase, numerous zymogen-like granules appear in the lower- and mid-apical regions and are released into the lumen. At this time extracellular digestion occurs in the stomach. In the excretory phase, the endoplasmic reticulum of the secretory cells becomes fragmentary and swollen, representing the end of protein synthesis and the reorganisation of organelles ready for the next feeding cycle.

Phasic activity of the digestive gland is correlated with the tidal cycle in *Littorina saxatilis* of high shore levels but not in *L. littorea* of mid- to low shore levels (Merdsoy and Farley, 1973). The significance of tidal synchrony is unclear, since although it is absent in *L. littorea* it occurs in *Rissoa parva* from extremely low levels on the shore (Wigham, 1976). The possible association of phasic activity of the digestive gland with foraging activity deserves further investigation.

3.2 Absorption

Food is absorbed at a rate equal to the product of the ingestion rate (section 2.3) and the absorption efficiency, the two factors often being interdependent. When passed more quickly through the gut, food is digested less thoroughly. But if food is plentiful, incomplete digestion may be unimportant because the extraction of nutrients is usually an asymptotic function of digestion time, governed by the 'law' of diminishing returns. This applies particularly to the more refractive foods, so herbivores and detritivores usually process food faster and in larger quantities than carnivores (section 2.3.2).

Absorption efficiency therefore depends on the digestion time and on the digestibility of the food. Animal tissue and bacteria are often more digestible than other foods. *Clione limacina* completely digests the tissues of *Spiratella retroversa*, absorbing over 90 per cent of the carbon and virtually 100 per cent of the nitrogen (Connover and Lalli, 1972); *Hydrobia ventrosa* has absorption efficiencies of 75 per cent for bacteria, 60-71 per cent for diatoms, 50 per cent for blue-green algae, and only 8 per cent for *Chroococcus*, a blue-green alga protected from digestive enzymes by a thick mucous coat (Kofoed, 1975a). In general, however, there is complete overlap between the absorption efficiencies of carnivorous and herbivorous species (Table 3.1). Interesting variations of absorption efficiency may occur among closely related herbivores. South African species of *Patella* graze micro-algae and red algal turf at different levels on the shore. Branch (1981) classified these limpets into 'exploiters' living among an abundance of food, which is processed rapidly in large quantities but with less efficiency, e.g. *Patella oculus* and *P. granatina* (absorption efficiency 72-79 per cent) and 'conservers' with restricted supplies of food that is processed rapidly but more efficiently, e.g. *P. granularis* (absorption efficiency 78-81 per cent), *P. longicosta* (86 per cent) and *P. cochlear* (93 per cent). Whereas *P. longicosta* and *P. cochlear* graze a sustained yield from a 'garden' of red algae (section 7.2.2), *P. granularis* lives high on the shore where algal growth is restricted.

The deposit-feeder *Hydrobia ventrosa* digests bacteria and diatoms efficiently (see above), but in common with other detritivores, much of its ingested material consists of non-living organic particles. *Hydrobia ventrosa* absorbs boiled hay, representing detritus, with an efficiency of 34 per cent, and although bacteria are absorbed with efficiencies of at least 70 per cent, the absorption efficiency for a mixture of bacteria and boiled hay is only 56 per cent (Kofoed, 1975a). Newell (1965) noticed that fine estuarine sediments often consist largely of faecal material, so the deposit-feeders must be consuming organic matter that has already been exploited by various animals. The enigma was solved by measuring the carbon and nitrogen content of faecal pellets both freshly produced by *Hydrobia ulvae* and after incubation in seawater. Evidently *H. ulvae* absorbed microbial protein from the detritus, so new faeces were low in nitrogen. Incubated faeces were recolonised by microbes that fixed nitrogen while using the detrital carbon as a substrate, making the faeces a rich source of nitrogen for deposit-feeders including *H. ulvae* itself.

Subsequently, it has been found that several detritus-feeders can absorb a significant proportion of their carbon requirements from the non-living detritus, but microbes are the main source of protein (Newell, 1983). At the present state of knowledge, however, generalisations require caution: one species, *Hydrobia totteni*, has been shown capable of subsisting entirely on micro-algae, apparently deriving little nutrition from bacteria even when these are abundant (Levington and Bianchi, 1981a).

Differential absorption of nutrients is of common, if not universal, occurrence. The nudibranch *Archidoris pseudoargus* feeds entirely on *Halichondria panicea* and related sponges, absorbing 58 per cent of the available carbohydrate and 93 per cent of the nitrogen. This highly efficient nitrogen absorption is

Table 3.1: Absorption Efficiencies of Gastropods, Defined as the Proportion of Ingested Energy Absorbed across the Gut Wall

Species	Food	Absorption efficiency (%)	Reference
Herbivores			
Patella vulgata	micro-algae	41	Wright and Hartnoll (1981)
Fissurella barbadensis	micro-algae	34	Hughes (1971b)
Nerita peloronta	micro-algae	42	Hughes (1971a)
N. tessellata	micro-algae	40	Hughes (1971a)
N. versicolor	micro-algae	39	Hughes (1971a)
Tegula funebralis	micro-algae	70	Paine (1971)
Littorina irrorata	micro-algae	45	Odum and Smalley (1959)
L. planaxis	*Ulva lactuca*	86	North (in Grahame, 1973)
L. obtusata	fucoids	73	Wright and Hartnoll (1981)
Carnivores			
Polinices alderi	*Tellina tenuis*	52	Ansell (1982)
Nucella lapillus	*Mytilus edulis*	66	Bayne and Scullard (1978a)
Archidoris pseudoargus	*Halichondria panicea*	52	Carefoot (1976b)
Navanax inermis	*Haminoea virescens*	50-70	Paine (1965)

Figure 3.5: Efficiencies with which Nudibranchs Absorb Amino Acids from their Food. A. Solid histogram, *Archidoris pseudoargus* feeding on *Halichondria panicea*. A specialist feeding only on this and closely related prey; absorption efficiencies are high and consistent. Dotted histogram, *Dendronotus frondosus* feeding on *Tubularia larynx*. A less specialised feeder taking a variety of prey; absorption efficiencies are extremely variable. After Carefoot (1967b). B. Solid histogram, *Aplysia punctata* feeding on the preferred alga in its natural diet, *Plocamium coccineum*. Absorption efficiencies are high and relatively even. Dotted histogram, *A. punctata* feeding on a less preferred alga, *Delesseria sanguina*; absorption efficiencies are more variable. After Carefoot (1967a)

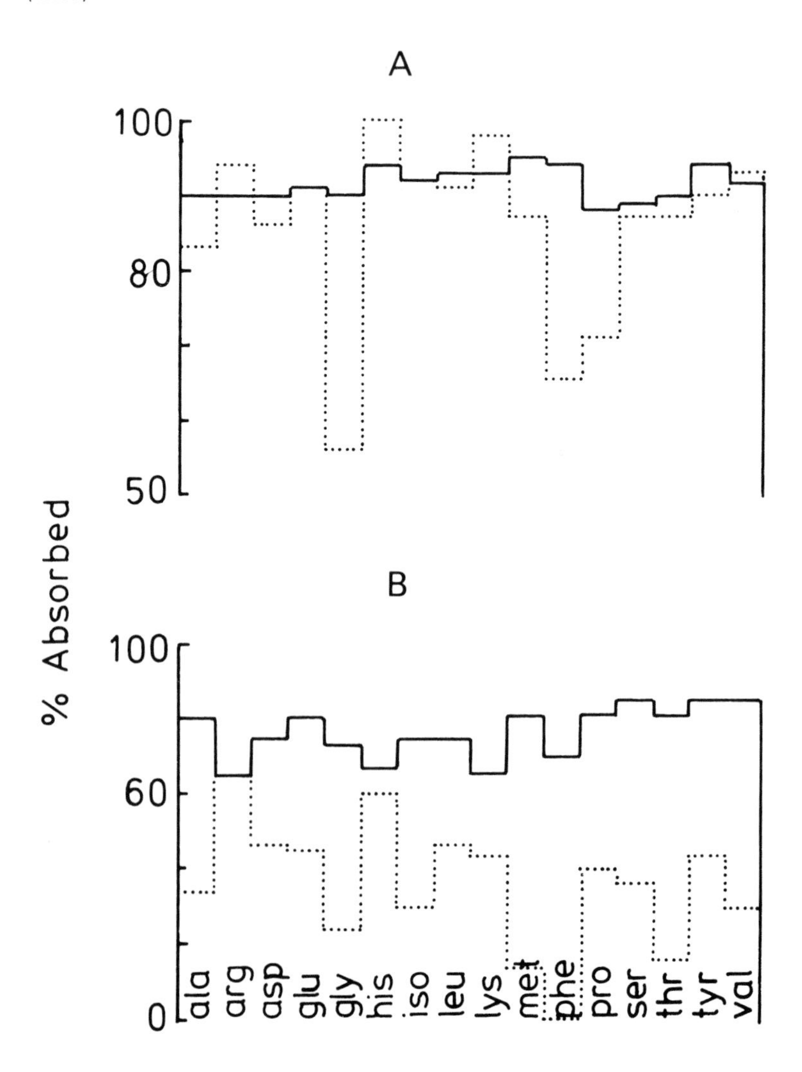

correlated with a uniformly high absorption of different amino acids (Figure 3.5A), reflecting the ability of *Archidoris* to meet its nutrititional requirements from a single species of prey (Carefoot, 1967b). It is similar to the high efficiency of nitrogen absorption by *Clione*, which also feeds on a single species of prey (see above).

Aplysia punctata has a less restricted diet and a less consistent pattern of absorption of amino acids from a single food species (Figure 3.5B). This trend is yet more pronounced in *Dendronotus frondosus*, which has an even wider diet (Figure 3.5A). Apparently, there is a counterbalancing decrease in the efficiencies of absorbing specific nutrients, associated with an increasing ability to exploit a greater variety of foods. Also, different kinds of food eaten by a heterophagous species tend to be absorbed with different efficiencies. *Aplysia punctata* absorbs amino acids more evenly and at a generally higher level of efficiency from its preferred food *Plocamium coccineum* than from the less preferred food *Delesseria sanguinea* (Figure 3.5B).

In addition to the quality and abundance of food, physiological state of the consumer may also be important: the absorption efficiency of *Polinices alderi* feeding on *Tellina tenuis* increases from 40 to 50 per cent when the snail enters its reproductive phase (Ansell, 1982). Hatchlings of *Planorbis contortus* absorb bacteria with an efficiency of 94 per cent, decreasing to 75 per cent throughout later stages of development (Calow, 1975). Veligers of *Ilyanassa obsoleta* feeding on several phytoplanktonic species are reported to have absorption efficiencies of 9-17 per cent (Pechenik and Fisher, 1979), much lower than values recorded in adult gastropods or in other planktivorous animals. Absorption efficiency does not appear to vary systematically with body size in post-larval stages or with temperature.

4 RESPIRATION, EXCRETION AND SECRETION

4.1 Introduction

Absorbed food provides the materials and energy for maintenance, synthesis and work. Metabolic energy used to power these processes is derived mainly from glucose, obtained directly from enzymatic transformation of other carbohydrates, fats and proteins in the food or stored in the body. Glucose, or a similar sugar, enters a series of enzyme reactions yielding energy that is stored, ready to do biochemical work, in the high-energy molecular configuration of ATP.

The first stage in these reactions, glycolysis, occurs in the cytoplasm and, since it does not involve oxygen, persists under conditions of anoxia (anaerobic respiration). The second, more efficient stage, the tricarboxylic acid cycle, occurs in the mitochondria and is an electron transport system using oxygen as the final electron acceptor. All materials used as a source of energy in normal aerobic respiration are ultimately broken down into carbon dioxide and water, with the addition of ammonia in the case of proteins, and are therefore unavailable for biochemical synthesis.

If all substances are measured in terms of their energy content, the amount of absorbed food used as a source of metabolic energy can be estimated from the rate of oxygen consumption and of nitrogenous excretion. The remaining absorbed food is either stored as lipid and glycogen or used in growth, reproduction and the production of external secretions. Growth and reproduction are dealt with in Chapters 5 and 6; external secretions of mucus and shell material are included with oxygen consumption and nitrogenous excretion in the present chapter because they represent energy used to support the animal rather than contributing to its productivity.

4.2 Intrinsic Determinants of Oxygen Consumption Rate

4.2.1 Body Size

Since the volume of metabolising cytoplasm is directly proportional to body weight, whereas gaseous exchange is limited by the area of respiratory surfaces, the rate of oxygen consumption of an animal might be expected to be proportional to (body weight)^b, where $b = ⅔ = 0.67$ (cf. p. 63). Gross comparisons among organisms ranging from protozoa, through invertebrates, to the largest mammals reveals a central tendency for b to equal about 0.75 (Hemmingsen, 1960). This departure from the 'surface law' has not been explained satisfactorily, but may involve increased respiratory efficiency brought about by the proliferation of organelles within the cell and by the development of complex respiratory and circulatory systems.

Respiratory and circulatory systems cannot be of particular importance, however, since excised tissues of *Littorina littorea* respire with the same weight-specific rate as the intact body; homogenisation of the tissues, on the other hand, obliterates this size dependency (Newell and Pye, 1971), suggesting the importance of internal cell membranes. Boddington (1978) proposes that the 'surface law' may not be involved at all, but that the metabolic rate of an organism is a function both of size and potential longevity, which together produce the allometric relationship between respiration rate and body size.

On the whole, marine gastropods conform approximately to the central tendency, but recorded values of b range almost from zero to unity (Table 4.1). Much of this variation is doubtless caused by experimental 'error', particularly as small samples have often been used, but some could be associated with size-specific changes in the physiology of the animals or in their response to environmental conditions. For example, metabolic changes associated with sexual processes in larger but not in smaller individuals would affect the value of b, as would different size-specific responses to temperature or to any other environmental variable. The value of b for *Littorina littorea* increases with the onset of warmer weather in spring (Newell and Pye, 1971). Larger individuals of *Patella granatina* respond less to changes in temperature than smaller ones, so b decreases as temperature rises (Branch, 1981). *Patella cochlear, P. oculus* and *P. granularis* all have b-values of about 0.71 when respiring in air but when under water they have mean b-values of 0.66, 0.73 and 0.80, respectively, paralleling their progressively higher zonation on the shore (Branch and Newell, 1978). Similarly, high- and low-level *P. vulgata* have b-values of 0.73 and 0.70, whereas the lower-shore *P. aspera* has a b-value of 0.65 (Davies, 1966). Perhaps in these cases, the respiration rate of

Table 4.1: Exponent (*b*) Relating Rate of Oxygen Consumption to Body Weight in Gastropods, Defined by the Power Function: $VO_2 = a\,(\text{Body Weight})^b$. Respiration is in water unless otherwise stated

Species	Exponent range	Exponent mean	Reference
Patella vulgata	0.51-0.79		Davies (1966)
P. aspera			
Patella spp. in air	0.46-0.86		Branch and Newell (1978)
Patella spp.	0.40-0.92		
Notoacmea pileopsis		0.64	Innes (1984)
N. pileopsis in air		0.72	Innes (1984)
Cellana radians		0.56	Innes (1984)
C. radians in air		0.65	Innes (1984)
C. tramoserica		0.71	Parry (1978)
Fissurella barbadensis	0.50-0.90	0.71	Hughes (1971b)
Nacella concinna		0.75	Ralph and Maxwell (1977)
Helcion pectunculus		0.43	Branch (1981)
Nerita tessellata		0.72	Lewis (1971)
N. versicolor		0.82	Lewis (1971)
N. peloronta		0.75	Lewis (1971)
Littorina littorea	0.03-0.60		Newell and Pye (1971)
Tegula funebralis	0.70-0.88		Paine (1971)
Polinices duplicatus	0.28-1.00	0.54	Huebner (1973)
P. alderi	0.54-0.69	0.63	Macé and Ansell (1982)
P. catena	0.55-0.70	0.70	Macé and Ansell (1982)
Thais lamellosa	0.19-1.02	0.60	Stickle (1973)
Nucella lapillus	0.34-0.96	0.51	Bayne and Scullard (1978a)
Nassarius reticulatus	0.42-0.95		Crisp *et al.* (1978)
Bullia digitalis		0.59	Brown and Da Silva (1979)
B. rhodostoma	0.62-0.80		Dye and McGwynne (1980)
B. digitalis	0.36-0.55		Dye and McGwynne (1980)
B. pura	0.21-0.44		Dye and McGwynne (1980)
Navanax inermis		0.89	Paine (1965)

animals from higher-shore levels is more size dependent than that of the lower-shore individuals. Such a trend has not been found in other species, however.

When invertebrates reach their physiological limits of lower thermal tolerance, metabolism may be suppressed sufficiently to obliterate its size dependency. Thus, the value of b approaches zero at 5°C in the cool-temperate snail *Polinices duplicatus* (Huebner, 1973) and at 20°C for tropical species of *Nerita* (Lewis, 1971).

Whatever its precise magnitude, a value of b less than unity implies a decreasing weight-specific rate of oxygen consumption as body size increases, specified by an exponent of $b - 1$. If $b = 0.75$, then the oxygen consumption rate per unit weight is proportional to (body weight)$^{-0.25}$. Smaller individuals therefore respire relatively faster than larger ones. This relationship is of particular interest when comparing species, for those with smaller bodies metabolise relatively faster, tending to have correspondingly shorter generation times and higher intrinsic rates of population increase than those with larger bodies.

4.2.2 Activity

Metabolism is commonly classified into three levels: standard (or basal) metabolism representing the energy required for maintenance; routine metabolism representing the energy used in synthetic and non-locomotory activity in addition to the standard metabolism; and active metabolism representing the energy used in locomotory or feeding activities (section 4.3.5) in addition to routine metabolism. Since it is impossible experimentally to control the synthetic and spontaneous muscular or ciliary activities of an intact animal, the true standard metabolic rate cannot be measured accurately. Although often labelled 'standard', the metabolic rate recorded when an animal lies quiescently in the respirometer is closer to the 'routine' level and is sometimes referred to as the 'resting' or 'quiescent' metabolic rate. The difference between the resting and active metabolic rate has, rather inappropriately, been called the 'scope for activity', but actually represents the energetic cost of the activity whether or not the full scope is realised.

When crawling, the rate of oxygen consumption of *Littorina littorea* is 1.39 ml 0^2 g^{-1} h^{-1}, falling to 0.17 ml 0^2 g^{-1} h^{-1} when the snails are quiescent (Newell and Pye, 1971). Crawling therefore uses an extra 1.22 ml 0^2 g^{-1} h^{-1}. Since the active is about eight times the resting metabolic rate, crawling appears to be a surprisingly costly activity for *L. littorea.* A high cost of crawling has also been reported for terrestrial slugs, estimated to use 0.90 J g^{-1} m^{-1} (Denny, 1980). These estimates are far higher than those for the cost of activity in other animals and may be inflated by the cost of mucus production (section 4.5). The South African whelk *Bullia digitalis* does not crawl by ciliary action over a mucous trail, as do littorines and terrestrial slugs, but pulls itself over the sand by a 'breaststroke' action of the foot and also burrows through the sand. Burrowing costs 0.25 J g^{-1} m^{-1} and crawling 0.15 J g^{-1} m^{-1}, comparable to the cost of running in terrestrial animals (Brown, 1982). *Bullia digitalis* also extends its foot to act as a 'sail' that is caught by the wave surge, transporting the whelk up the beach on the flooding tide. While 'surfing', *B. digitalis* keeps the foot turgid and twists it violently from side to side. A surfing whelk uses oxygen 2.24 times faster than when lying quiescently beneath the sand, whereas a crawling whelk uses only 1.22 times as much oxygen; but as surfing carries the whelk 10 m or more in a few seconds, it is much cheaper than crawling up the beach.

Sedentary filter-feeders (section 2.1.4) do not normally incur costs of locomotion, but when feeding they spend extra energy on mucus secretion, postural movements and the generation of a strong inhalant current. By simultaneously measuring the rates of oxygen consumption and clearance of suspended algal cells from the seawater, Newell and Kofoed (1977) estimated the cost of filtering in *Crepidula fornicata.* In experimentally filtered seawater, *C. fornicata* used 0.36 ml 0_2 g^{-1} h^{-1} and an extra 0.44 ml 0_2 g^{-1} h^{-1} when filtering suspended diatoms. Therefore *C. fornicata* uses 2.3 times more oxygen when filtering than when merely generating a respiratory current, comparable with the cost of locomotion in *Bullia digitalis.* Increased pumping rate, like increased crawling speed, will use energy faster (Figure 2.17B) and so it is to be expected that the pumping rate will be adjusted to the density of suspended food, giving the maximum yield per unit cost (section 2.2.2).

4.2.3 Endogenous Rhythms

Endogenous activity rhythms are widespread among animals, and in the case of intertidal species the periodicity of the rhythms may be diurnal or tidal. When kept under constant laboratory conditions, *Littorina nigrolineata* continues to show peaks of crawling activity phased with the concurrent tides (Petpiroon and Morgan, 1983). Oxygen consumption and rhythmic activity have not been measured simultaneously in any intertidal gastropod, but a correlation between them is to be expected. Diurnal and tidal rhythms in oxygen consumption have been detected in *L. littorea* (Sandeen *et al.*, 1954) and diurnal respiratory rhythms in submerged *L. saxatilis* (Sandison, 1966). Endogenous rhythmicity ensures that animals respond quickly and unerringly to cyclical changes in environmental conditions without being stimulated inappropriately by unpredictable short-term fluctuations (Naylor, 1976). The proximate significance of activity rhythms, however, is not always apparent. *Littorina irrorata* responds to nocturnal darkness by an increased rate of oxygen consumption (Figure 4.1), but under experimental conditions this response is not correlated with activity levels. In the field, nocturnal activity might reduce the risk of predation.

4.2.4 Growth and Reproduction

The weight-specific rate of oxygen consumption of most invertebrates fluctuates seasonally. This is generally regarded as a response to seasonal changes in temperature, but Parry (1978) suggested that, at least in the Australian limpet *Cellana tramoserica*, it may be the direct effect of seasonal patterns of somatic

Figure 4.1: The Mean Respiration Rate of *Littorina irrorata* kept in the Laboratory at 20°C under Alternating Light and Dark Periods (but Otherwise Constant Conditions) Cycles Continually, with the Onset of Peaks Slightly Anticipating the Periods of Darkness. The endogenous circadian rhythm persists even though the mean level of respiration rate declines during captivity. After Shirley and Findley (1978)

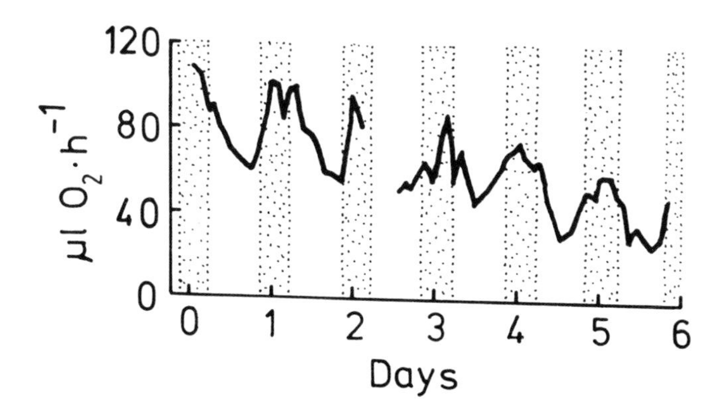

and gonadal growth and only indirectly correlated with changes in temperature. Since growth is too slow to be measured in the short time periods used for respirometry, Parry measured monthly increments of the shell. Knowing the relationship between shell length and body weight, he calculated instantaneous growth rate (cal h^{-1}) for each month. By assuming that maintenance requirements remained constant and that net growth efficiency (assimilated energy/energy accumulated in growth) was 67 per cent, he predicted the respiration rates necessary to account for the estimated growth rates. The predicted respiration rates closely matched the observed rates (Figure 4.2), suggesting that growth and not temperature determined seasonal patterns of respiration in *Cellana tramoserica.* This interpretation was corroborated by partial suppression of the seasonal increase in respiration rate when snails were caged at high densities, causing them to deplete their food supply (Parry, 1984). The starving snails had insufficient food to grow and their respiration rate remained low during the normal growth season. However, during the reproductive season, gonadal development proceeded in spite of starvation and the respiration rate increased to normal levels for the time of year (Figure 4.2).

Tropical animals tend to grow faster, with shorter lifespans, than equivalent temperate species, and this may be associated with a higher metabolic rate. *Bullia melanoides* on tropical Indian beaches at 30°C has a routine respiration rate about 10 times

Figure 4.2: Seasonal Changes in Growth Rate Alter the Respiration Rate of *Cellana tramoserica* Independently of Temperature. The curve represents predicted respiration rate at 15°C, based on a knowledge of growth rate, and the circles represent observed respiration rates after two days at 15°C in the laboratory at different times of the year. Diamonds denote the respiration rate of snails kept short of food. During June-August, these snails do not grow and have a respiration rate close to that in the non-growing season, In October, however, the starved snails undergo gonadal development and this is accompanied by an elevated respiration rate. The data represent a 'standard' limpet of 0.25 g dry body weight. After Parry (1978, 1984)

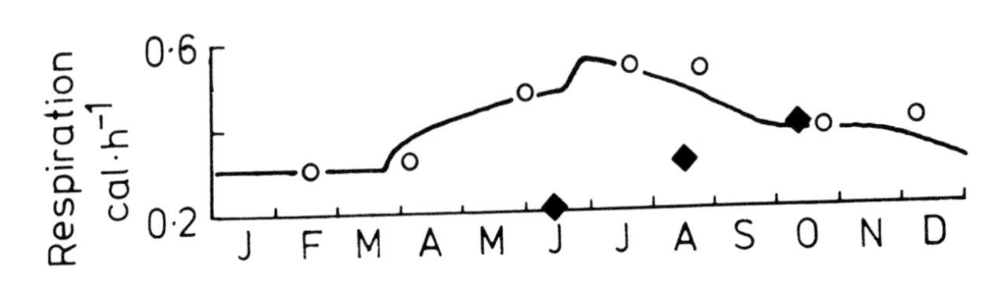

greater than that of *B. digitalis* on South African beaches at 15°C (Brown *et al.*, 1978). Growth and reproductive rates are less in *B. digitalis* than in *B. malanoides* but their proportional contribution to the difference in metabolic rate is unknown. Measuring the metabolic cost of growth and reproduction is a difficult but important task for future research.

4.3 Extrinsic Determinants of Oxygen Consumption Rate

4.3.1 Temperature

The mean rate of oxygen consumption of animals held in respirometers is frequently an increasing function of temperature. The factor of increase in respiration rate over a 10°C rise in temperature (Q_{10}) tends towards 2, similar to the reaction rate of enzyme systems. Newell and Northcroft (1967) however, recorded the respiration rate of *Littorina littorea* continuously, rather than taking a mean value for the entire experimental period. In doing so, they noticed that whereas the respiration rate of crawling snails was sensitive to temperature, the respiration rate during periods of quiescence was not ($Q_{10} = 1$). This was an exciting discovery since it meant that energy would be saved at high temperatures whereas standard metabolism would persist at a functional rate at low temperatures.

Temperature independence is not, however, a universal feature of quiescent metabolic rate. For example, that of *Patella vulgata* and *P. caerulea* increases with temperature (Davies and Tribe, 1969). On the other hand, the routine metabolic rate (section 4.2.2) of *Bullia digitalis* is independent of temperature between 10 and 25°C, but positively correlated with temperatures on either side of the range (Figure 4.3). This flattening of the rate-temperature curve will enable *B. digitalis* to conserve energy, as ambient temperatures on the beach can fluctuate from 8 to 17°C within a tidal cycle (Brown, 1982). Hence *B. digitalis* is metabolically suited to the widely fluctuating temperature regime on the west coast of the Cape Peninsula, South Africa, but on the east coast, where temperatures fluctuate less, the related species *Bullia rhodostoma* does not show metabolic independence of temperature (Brown and Da Silva, 1984).

Whereas standard or even routine metabolic rate may be

Figure 4.3: The Respiration Rate of *Bullia digitalis* at Moderate Levels of Activity is Independent of Temperature between the Limits of 10 and 25°C Normally Experienced in its Habitat. The graph represents a snail of 750 mg dry weight. After Brown and Da Silva (1979)

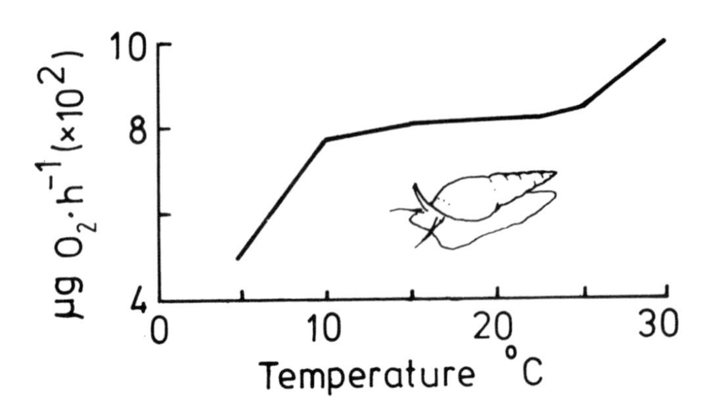

independent of temperature in some gastropods positive temperature dependence is widespread and is always the case with active metabolic rate. Animals may exploit this temperature dependence to increase activity levels at high temperatures, or they may dampen it (acclimate) to conserve energy at high temperatures and maintain adequate metabolic rates at low temperatures.

Acclimation, the tendency to maintain a uniform metabolic rate over prolonged changes in environmental temperatures, is achieved by lateral translation of the rate–temperature curve, by clockwise rotation of the curve, or by a combination of the two (Prosser, 1973). The biochemical mechanisms of acclimation are not fully understood, but may include changes in the concentration of specific enzymes, changes in the types of enzymes synthesised (Figure 4.4), or changes in the structure of enzymes after synthesis. *Crepidula fornicata* acclimates by translation and *Littorina littorea* by translation and rotation of the rate–temperature curve (Figure 4.5).

Whereas translation maintains a constant metabolic rate over gradual changes in temperature, clockwise rotation dampens the responsiveness to immediate fluctuations in temperature. *Crepidula fornicata* experiences gradual seasonal changes in temperature but is protected somewhat by its low-intertidal habitat from tidal fluctuations in temperature. *Littorina littorea*, on the

Figure 4.4: Enzyme Polymorphism May Be Involved in Temperature Acclimation in *Acmaea limatula*. The activity of malic dehydrogenase was tested in limpets acclimated to 8°C and 18°C. Activity was expressed as the concentration (K_m) of substrate (oxalacetate) equivalent to the reciprocal of the binding coefficient in the enzyme-substrate reaction. The minimum in the activity-temperature curve represents the point of maximum affinity of the enzyme for the substrate. Since the minima for limpets acclimated to 8°C and 18°C were different, it was concluded that at least two enzymatic proteins occur, which vary in proportions that are determined by environmental temperature. After Markel (1976)

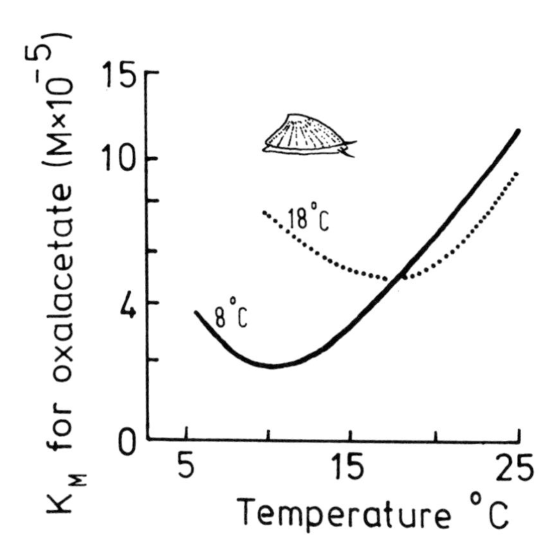

other hand, experiences longer periods of tidal emersion during which the air temperature may contrast sharply with that of the sea. Clockwise rotation of the rate–temperature curve will therefore buffer the routine metabolic rate of *L. littorea*, conserving energy in a strongly fluctuating temperature regime.

Counter-clockwise rotation of the rate–temperature curve (negative acclimation) signifies accentuated metabolic sensitivity to temperature and occurs during winter in *Bullia pura*, allowing it to metabolise rapidly during warm spells (Dye and McGwynne, 1980). Negative acclimation also occurs in *Polinices duplicatus* living on the climatically variable mudflats of the New England coast (Huebner, 1973). *P. duplicatus* has a rather flat rate–temperature curve with Q_{10} values below 2 over the normal temperature range throughout most of the year, conserving energy when temperatures fluctuate. During November and December, however, the Q_{10} rises above 2 between 5-10°C, so maintaining an

Figure 4.5: A. Left: Rate-Temperature Curves for the Routine Respiration Rate of *Crepidula fornicata* of Standard 160 mg Dry Body Weight previously Acclimated at 10, 15, and 25°C. Acclimation shifts the curves along the abscissa (lateral translation). Right: the routine respiration rate at the acclimation temperatures is almost constant. After Newell (1979) (from Newell and Kofoed, 1977). B. Rate-Temperature Curves for the Routine Respiration Rate of *Littorina littorea* Collected from the Shore between Winter and Summer when the Environmental (Acclimation) Temperatures were 14, 17, and 30°C. Acclimation shifts the curves along the abscissa (lateral translation) and also rotates them towards the horizontal at higher temperatures. After Newell (1979) (from Newell and Pye, 1971)

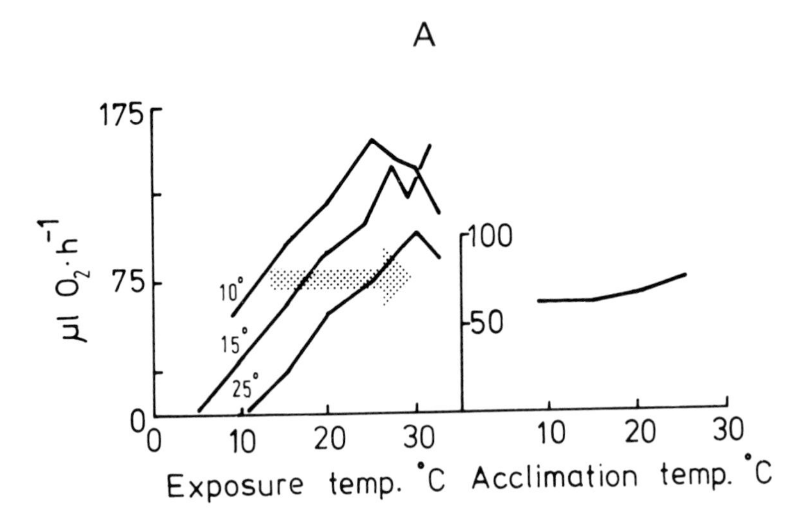

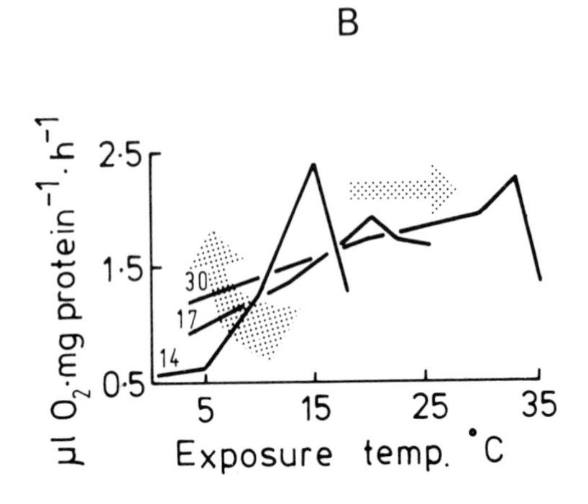

adequate metabolic rate. This trend is even more pronounced in *Nucella lapillus*, which has low Q_{10} values of 1.2-2.0 during summer, when it is therefore relatively insensitive to fluctuations in temperature, but develops Q_{10} values of 2.0-6.5 in winter and so is able to respond quickly to rising temperatures in the spring (Bayne and Scullard, 1978a). In contrast, the largely subtidal species *Polinices alderi* and *P. catena* have Q_{10} values of about 2 at all normal environmental temperatures and this lack of independence is to be expected among species that do not experience wide fluctuations in temperature.

4.3.2 Oxygen Tension

Shortage of oxygen is likely to be encountered by subtidal gastropods that burrow into sediments and by intertidal species that withdraw into their shells when desiccated at low tide. Gastropods that habitually experience low oxygen tensions might be expected to be 'oxy-regulators', increasing the efficiency of uptake as the oxygen supply decreases, whereas low-shore and surface-dwelling subtidal species would not be expected to do this, being 'oxyconformers'. Such is the case among chitons, close relatives of the gastropods, in which aerial respiration is maintained over large decreases in oxygen tension in high-shore but not in low-shore species (Murdoch and Shumway, 1980). Since oxygen enters the tissues by passive diffusion, 'oxyregulation' must involve physical changes of the respiratory system to facilitate oxygen uptake, but it is not known what these might be. Most gastropods are, however, likely to be oxyconformers (Figure 4.6).

Prolonged anoxia will enforce anaerobic respiration and this may be of regular occurrence in some gastropods. Once oxygen becomes available again, the products of anaerobic respiration can be metabolised aerobically and the amount of oxygen required for this is the 'oxygen debt'. *Littorina neritoides* extends to the splash zone on exposed rocky shores, where it may be kept dry for several weeks during calm, fine weather. Under such conditions *L. neritoides* closes the shell aperture with its operculum, remaining quiescent until wetted. Oxygen diffusion around the edges of the operculum must be minimal, and it seems likely that anaerobic respiration will be important during aestivation, although this has not been tested.

When experimentally subjected to several hours of 'environmental anoxia', *Nassarius mutabilis* breaks down the substrates

Figure 4.6: An Oxyconformer *Nassarius reticulatus* Decreases its Rate of Oxygen Uptake as Oxygen is Depleted from the Water. The relative oxygen uptake is expressed as a percentage of the uptake at 80% saturation. After Crisp *et al.* (1978)

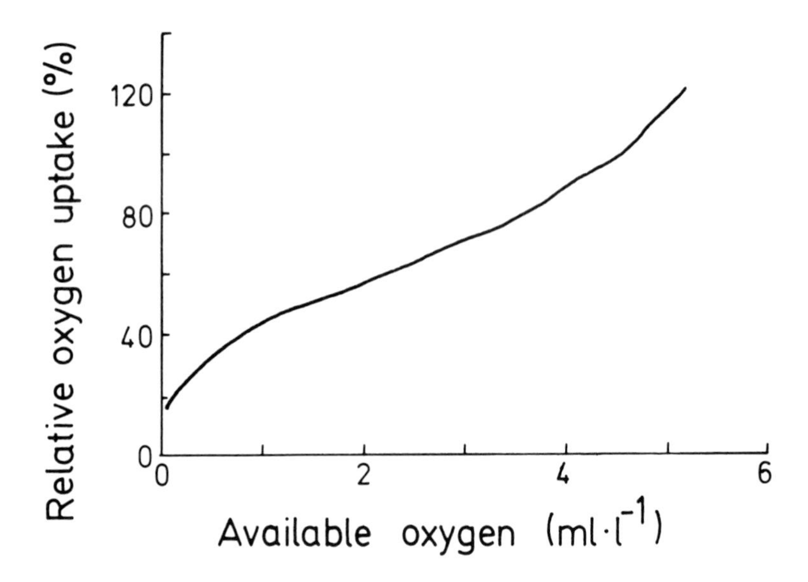

phosphoarginine, glycogen and aspartate to form alanine and succinate (Gäde *et al.*, 1984). *Bullia digitalis* also produces alanine under similar conditions (Brown, 1982). *Morula granulata* breaks down glycogen and to a lesser extent aspartate to form succinate and alanine when immersed in water of less than 15 per cent salinity (Uma Devi *et al.*, 1984). To avoid osmotic imbalance, the snail closes the operculum across the shell aperture, isolating itself from the surrounding water. The operculum closes more tightly as salinity decreases below 15 per cent and this is accompanied by a greater predominance of anaerobic respiration, reflected by a greater oxygen debt, as gaseous exchange is reduced. Compared with the classical glycolytic pathway producing lactate, the succinate pathway has a higher ATP output per mole of glucose used.

Temporary anoxia occurs in muscles during sudden, strenuous exercise, and when this 'functional' or 'physiological' anoxia occurs in *Nassarius mutabilis*, arginine phosphate and glycogen are broken down to form octopine, as is the case among other invertebrates. Formation of octopine, or related products, apparently does not increase the yield of ATP per mole of glucose, but perhaps has other advantages such as stabilising the acid-base balance of the cell or preserving the regulatory properties of enzymes in the cell (Fields, 1983).

Thermodynamic considerations suggest that the maximum rate and maximum efficiency of anaerobic metabolism are mutually exclusive (Gnaiger, 1983). During physiological anoxia, energy is needed quickly because of the associated intense muscular activity, and thermodynamically grossly inefficient pathways sustain high metabolic rates for brief periods. But under long-term environmental anoxia a low, steady metabolic rate is achieved by thermodynamically more efficient pathways. These pathways are probably switched into operation by cellular changes of pH and adenylate phosphorylation potential associated with prolonged anoxia. Moreover, calorimetric studies directly measuring metabolic heat output indicate that unknown sources of energy may be important among marine invertebrates experiencing environmental anoxia.

4.3.3 Salinity

Estuarine and intertidal gastropods experience cyclical or intermittent salinity fluctuations either in the ambient seawater or in the mantle-cavity fluid (section 1.6), but since they are largely osmoconformers, their metabolic rate would not be expected to increase above normal when salinities fluctuate. On the contrary, the respiration rate of *Thais haemastoma* falls in response to lowered salinity and, since accompanied by withdrawal of the siphon, it is probably caused by decreased ventilation (Findley *et al.*, 1978). Similarly, many other intertidal gastropods, including *Littorina littorea*, withdraw into their shells when exposed to low salinities, with a concomitant cessation of oxygen consumption (Shumway, 1978). Anaerobic metabolism must be enforced during these conditions, the end products being metabolised aerobically when salinity returns to normal.

4.3.4 Respiration in Air and Water

Subtidal gastropods respire across exposed epithelia, especially over the gill which generates a respiratory current. Exposure to air causes withdrawal into the shell, collapsing of the gill and a decreased rate of respiration. Intertidal species maintain their oxygen consumption in air by ctenidial (branchial) respiration in the mantle-cavity fluid and by making greater use of vascularised mantle tissue (section 1.6). Ctenidial respiration is more efficient in water, and diffusion across the mantle is more efficient in air.

The difference between the rate of aerial and aquatic respiration varies among intertidal species, those from lower levels on the shore tending to have lower and those from higher levels greater aerial than aquatic respiration rates. Pairs of lower-shore and higher-shore species conforming to this trend include *Patella caerulea* and *P. lusitanica* (Bannister, 1974), *Notoacmea fenestrata* and *Collisella digitalis* (Doran and McKenzie, 1972), and *Lacuna vincta* and *Littorina littorea* (McMahon and Russell-Hunter, 1977). Exceptions do occur: West Indian species of *Nerita* from lower and higher shore levels have similar respiration rates in air and water (Hughes, 1971a).

Morphology of the respiratory organs may affect the differential effects of submersion and emersion on respiration rate. The high-shore limpets *Notoacmea pileopsis* and *Cellana radians* respire slower in air than in water, but the effect of desiccation is greater on *N. pileopsis*, which has a bipectinate gill (section 1.7), than on *C. radians*, which has secondary pallial gills (Innes, 1984).

An interesting, more complex pattern of variation in aerial and aquatic respiration rates occurs among South African limpets (Branch, 1981). *Patella cochlear* from the lowest shore levels has equal aquatic and aerial respiration rates. Mid-shore *P. oculus* and mid- to high-shore *P. granularis* vary in their response to aerial exposure according to body size, but in different ways. Small individuals of *P. oculus* live in rock pools and damp crevices, respiring faster in water than in air, whereas large individuals occur on dry rocks and respire faster in air (Figure 4.7). Metabolic rate is therefore kept high by a reversal of the aerial/aquatic respiratory response as the limpets age, enabling them to maintain a high productivity by rapidly exploiting the plentiful food at mid-shore levels.

On the other hand, *P. granularis* migrates up the shore as it ages, young individuals occurring lower on the shore and respiring faster in air than in water, older and therefore larger individuals occurring high on the shore but respiring faster in water than in air. This response to exposure and wetting is contrary to the general trend among high intertidal gastropods but, since tidal immersion is brief at these high levels on the shore, it will have the effect of lowering the average metabolic rate. Conservation of energy is

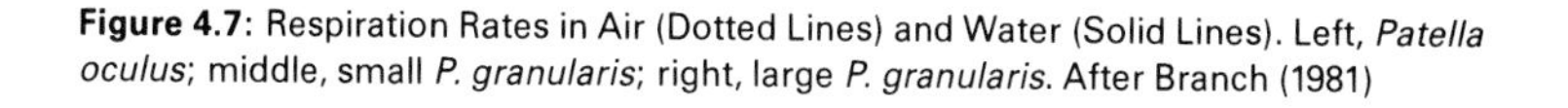

Figure 4.7: Respiration Rates in Air (Dotted Lines) and Water (Solid Lines). Left, *Patella oculus*; middle, small *P. granularis*; right, large *P. granularis*. After Branch (1981)

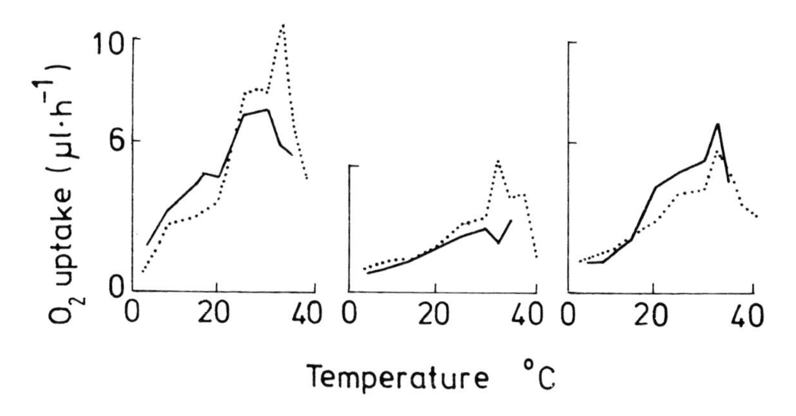

probably important to *P. granularis* because food is scarce at the top of the shore.

4.3.5 Food Supply

Food intake, or lack of it, can profoundly influence metabolic rate. An increase in metabolic rate immediately following the consumption of food has been noted in a variety of animals and is often referred to as the 'specific dynamic action', perhaps associated with a surge of catabolism or anabolism. When given a meal of crab flesh after a period of starvation, *Buccinum undatum* excretes a pulse of ammonia during the following 24 h, signifying the deamination of absorbed amino acids (Figure 4.10A). There is a concomitant doubling of the rate of oxygen consumption, but this elevation is maintained at least for 50 h, so is unlikely to represent only the metabolic cost of protein catabolism.

Specific dynamic action is not usually distinguishable from other causes of increased respiration accompanying feeding. The rate of oxygen consumption of *Nucella lapillus* begins to rise while the dogwhelk drills its prey, continuing to a peak during ingestion (Figure 4.8). Although part of this respiratory increase must be associated with the metabolic costs of drilling and ingestion, some of it may be attributable to metabolic stimulation caused by the presence of food.

Gastropods are extremely sensitive to proteins emanating from their food: *Ilyanassa obsoleta* responds to a 10^{-10} M dilution of

Figure 4.8: When a Recently Fed *Nucella lapillus* is Deprived of Food, its Respiration Rate Declines until the Next Meal and then Rises Quickly to its Original Value during Drilling and Ingestion. The curve represents the respiration of a snail of 80 mg dry flesh weight. After Bayne and Scullard (1978b)

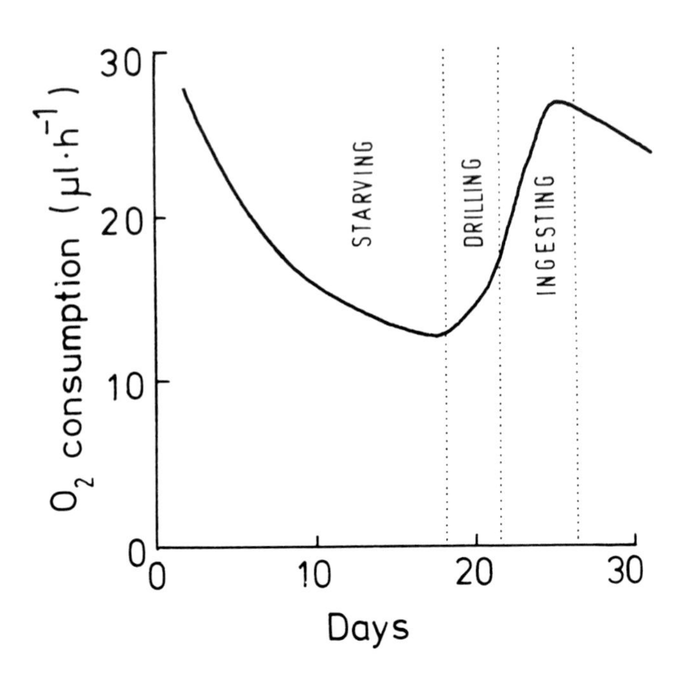

water-soluble glycoprotein, MW 120000, isolated from oysters (Gurin and Carr, 1971) and *Elysia cauze* responds to a 5×10^{-8} M solution of proteins isolated from the alga *Caulerpa* (Jensen, cited by Kohn, 1983). The mud-snail *Nassarius reticulatus* responds quickly to the scent of macerated crab or even of glycine. Within minutes, the rate of oxygen consumption rises 3-4 times above the routine level, sometimes, but not always, accompanied by increased activity. Recently fed snails show the respiratory response without everting their proboscis or showing any other signs of increased activity. When the olfactory stimuli are removed, the respiration rate quickly falls to the routine level. It also declines after 15-30 min of continuous exposure to the olfactory stimuli (Figure 4.9A). This anticipatory increase of metabolic rate in response to olfaction may reflect physiological preparations of the digestive system, particularly enzyme production, in readiness for a meal, although this interpretation requires substantiation. When allowed to feed, the respiration rate of *N. reticulatus* gradually declines to the routine level over 3 or 4 days (Figure 4.9B). This

Figure 4.9: A. The Smell of Food Increases the Rate of Oxygen Consumption by *Nassarius reticulatus*. Respiration rate per unit dry body weight declines after the meal. The arrow indicates the time when food was introduced. The dotted lines indicate the beginning and end of feeding. B. After a Meal, the Rate of Oxygen Consumption per Unit Dry Body Weight Declines over Several Days towards the 'Standard' Rate. The arrow indicates the time when food was introduced and eaten. After Crisp *et al.* (1978)

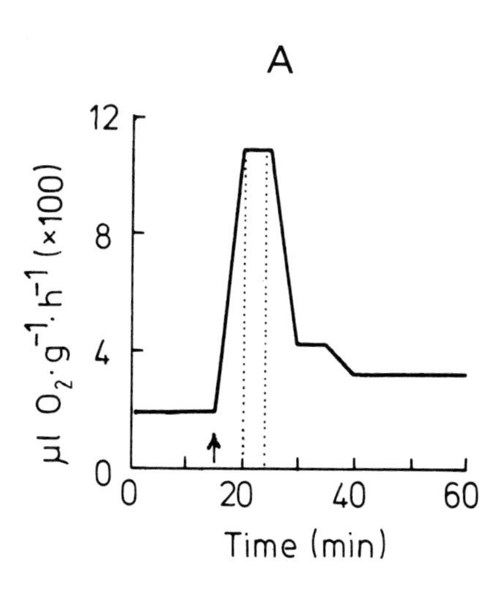

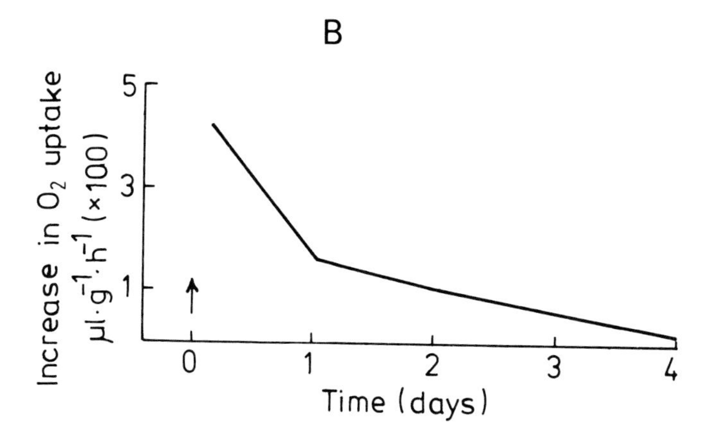

extended phase of elevated oxygen consumption would include any 'specific dynamic action'.

Following the increased levels associated with a meal, the oxygen consumption rate of *Nucella lapillus* continues to decline until the next meal, at least for up to 18 days (Figure 4.8). Periods of this length without food are commonly endured by *N. lapillus* when wave action or low temperature prevent foraging, the declining metabolic rate being a means of saving energy. However, the related species *Nucella lamellosa* shows a tendency to increase its rate of oxygen consumption when starved for longer periods of up to 53 days (Stickle, 1971). Prolonged starvation may induce greater activity, so increasing the area searched for prey.

4.4 Nitrogenous Excretion

The nitrogenous end product of protein metabolism in marine gastropods is mostly ammonia (Figure 4.10A), much of which diffuses across external epithelia. In so doing, ammonia raises the pH of mantle epithelium and probably helps the process of shell secretion by enhancing the conversion of soluble bicarbonate into insoluble carbonate ions. Urea is not usually produced, but uric acid occurs in the tissues in amounts proportional to the body mass and is perhaps used as a depot of nitrogen rather than as a dump for waste material (Duerr, 1968).

Littorina littorea, however, excretes uric acid and some urea when exposed to the air at low tide, reverting to the excretion of ammonia when immersed by the rising tide (Daguzan and Razet, 1971). Excretion of nitrogen as uric acid or urea is metabolically more expensive than producing ammonia, but conserves water. The alternation between forms of nitrogenous excretion that are typical of terrestrial and aquatic animals during periods of tidal emersion and immersion may prove to be widespread among intertidal gastropods.

The rate of ammonia excretion has an allometric relationship with body size (Figure 4.10B), similar to the rate of oxygen consumption (section 4.2.1). Although accounting for a proportion of the energy used in metabolism, nitrogenous excretion has rarely been included in the measurement of energy budgets. Technical difficulties of measuring nitrogenous excretion are greater than those of respirometry: small quantities of nitrogen are pro-

Figure 4.10: A. A Pulse of Ammonia Excretion by *Buccinum undatum* Occurs Immediately after Feeding. The rate of ammonia excretion per unit dry body weight then gradually declines towards the 'starvation' level (dashed line). After Crisp *et al.* (1981). B. The Rate of Ammonia Excretion is Allometrically Related to Body Weight, Giving a Straight Line on a Double Logarithmic Plot. The slope of the regression line is 0.59 for *Calliostoma ligatum*, 0.57 for *Littorina sitkana* and 0.63 for *Nucella lamellosa*. After Duerr (1968)

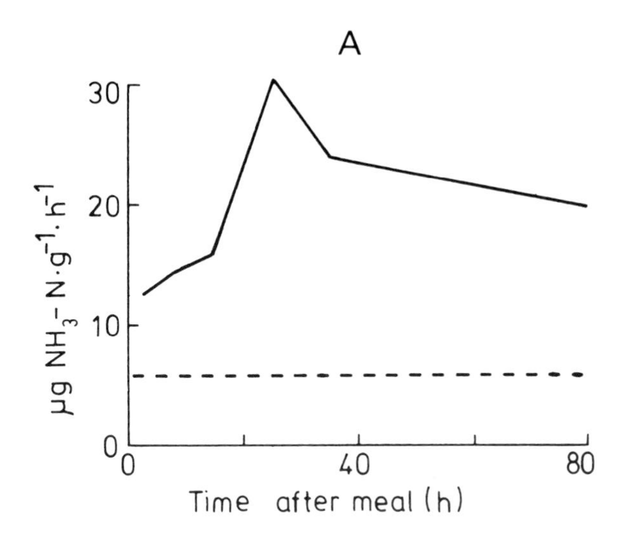

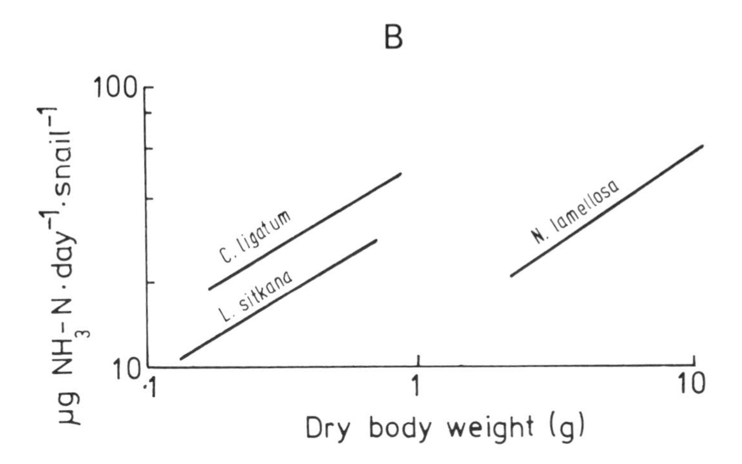

duced and are easily confounded with extraneous bacterial nitrogen, and some nitrogen is excreted with the faeces. Wright and Hartnoll (1981), however, included ammonia excretion when estimating the energy budget of *Patella vulgata* but found it to be an insignificant factor compared with respiration.

Nitrogenous excretion can be strongly affected by physiological condition. Predatory gastropods have a notorious ability to survive long periods of starvation, enabling them to cope with unpredictable supplies of food, and, at least in the case of *Nucella lapillus*, almost all their energy requirements are met from protein catabolism. Ammonia excretion by *N. lapillus* increases throughout starvation, with losses of primary amines building up to a peak during the first three weeks and declining to control levels thereafter (Stickle and Bayne, 1982). Subnormal salinities, towards the lower physiological tolerance of the animal, interact with nitrogenous excretion. At 17 per cent salinity, the loss of primary amines accounts for a significant proportion of the excreted nitrogen by *N. lapillus* and is greater than at all other salinities. Ammonia excretion, on the other hand, fluctuates but shows no systematic trend with changing salinities.

4.5 Mucus Secretion

All gastropods secrete mucus for feeding, locomotion or protection (section 1.8). That used to lubricate the alimentary canal may be partly reabsorbed, although some is always expelled with the faeces. In the freshwater limpet *Ancylus fluviatilis*, faecal mucus may be as much as 4-6 per cent of the ingested energy (Calow, 1974). Filter-feeders ingest their mucous webs, but some wastage may occur (section 2.1.4). Cleansing of the mantle cavity by ciliary rejection of mucous-bound particles (pseudofaeces) is a potential source of loss, although in vermetids and vermiculariids the pseudofaeces are often eaten (Hughes, 1985b).

The main loss of mucus is in trails used for locomotion, amounting to 9 per cent of the ingested energy of *Ancylus fluviatilis* (Calow, 1974), about 8 per cent of absorbed energy in the territorial limpet *Patella longicosta* (Branch, 1981) and about 9 per cent of the absorbed energy of *Hydrobia ventrosa* (Kofoed, 1975b); none of these figures included the metabolic cost of mucus secretion itself.

This energy expenditure must, of course, buy dividends in the form of efficient adherence (Grènon and Walker, 1980) and locomotion (Denny, 1980). Moreover, in gastropods that move within home ranges or feed in aggregations, mucous trails can be an important source of directional information. The slug *Onchidium verruculatum* emerges from crevices on the ebbing tide to graze over the rocks, eventually retracing the outward trail back to the home crevice. Rotating a segment of the trail, made on an artificial substratum, through 180° causes the returning slug to move away from its home, proving that directional information is present in the mucus. To be of use as guides to homes within overlapping foraging areas, trails must have individual identity and, indeed, slugs follow only their own trails, presumably identified by chemical information (McFarlane, 1981).

Littorina planaxis are able to detect the polarity of their trails, probably using sense organs at the tips of the cephalic tentacles (Raftery, 1983). The snails can still detect the polarity when a 100 μm mesh is placed over the trail, but not when the mesh is reduced to 0.22 μm. Mud-snails *Ilyanassa obsoleta* follow each other's trails, always heading towards the front of the trail. Polarity of the trail is probably detected from structural elements in the mucus; staining reveals parallel and interconnecting filaments, and trail-following is abolished if the mucus is denatured by heating to 80°C (Bretz and Dimock, 1983).

Whereas trail-following causes the formation of feeding aggregations in *Ilyanassa obsoleta*, it can result in the formation of protective aggregations in other species. On tropical Indo-Pacific shores, *Nerita textilis* graze the damp rock on the falling tide, but as the substratum dries out they follow each other's trails when these happen to cross. If a snail finds a hollow in the rock, it stops moving and others following its trail aggregate in the same hollow (Focardi *et al.*, 1985). The aggregations reduce desiccation and overheating.

On the Pacific coast of Panama the same mechanism causes *Nerita scabricosta* to aggregate in crevices and tide pools. Aggregations contain up to several hundred snails and most consist of two or more layers with the larger individuals on top. Clustering helps to trap moisture leaking from the mantle cavity and reduces the rate of mortality, especially of smaller snails (Garrity and Levings, 1984). Aggregation in response to desiccation has also been described for the Australian muricid *Morula marginalba*

(Moran, 1985), and on South African shores, *Nodilittorina africana knysnaensis* sometimes forms clusters, perhaps with a similar function (Plate 4.1).

Nucella lapillus forms dense aggregations in winter at low shore levels and, by reducing the total surface area per unit mass, this probably reduces the risk of being dislodged by severe wave action (Feare, 1971a). They also aggregate to lay eggs in the spring (section 6.5.1), but whether *N. lapillus* uses mucous trails in its aggregating behaviour remains to be established.

On sheltered shores, the mucous trails of gastropods are often visible at low tide on the surface of stones, where silt adhering to the mucus becomes pale grey as it dries out. Not only does mucus trap silt, but it also traps micro-organisms which may thrive and even exploit the mucus as a substratum or as a source of carbon.

Enriched by the growth of micro-organisms, mucus becomes a potentially valuable food for grazers and deposit-feeders (Calow, 1979). On the exposed shores of western North America the limpets *Collisella scabra* and *Lottia gigantia* each forage within a restricted home range centred on the 'scar' upon which they sit

Plate 4.1: Aggregation of *Nodilittorina africana knysnaensis* Form on Horizontal Surfaces when Emersed for Prolonged Periods by Neap Tides. Exposed to intense isolation and a drying atmosphere, the aggregations reduce the total surface area per unit mass and so may alleviate overheating and desiccation. The behavioural mechanisms causing aggregation are unknown. Each snail is about 10 mm in height

when at rest. Within a home range, mucous trails cross one another extensively and the occupant limpet frequently retraces its own path. In doing so it ingests the old mucus, which has been experimentally shown to trap and stimulate growth of micro-algae (Connor and Quinn, 1984). Another limpet, *Collisella digitalis*, forages more widely without returning to a scar and therefore seldom retraces its own path. Moreover, *C. digitalis* tends to form dense aggregations in which different individuals are likely to graze each other's trails. Accordingly, the mucus of *C. digitalis* does not stimulate the growth of micro-algae.

Among intertidal gastropods, mucus sometimes plays a key role in preventing excessive desiccation. Species of winkle, such as *Littorina saxatilis* and *L. neritoides*, living high on the shore, secrete a mucous membrane across the shell aperture when neap tides or calm dry weather leave them unwetted for long periods. The mucus also cements these snails to the rock, holding them secure even though the foot is deeply withdrawn into the shell.

A similar anti-desiccation device is well developed in certain acmeid limpets such as *Collisella digitalis*, which secrete a curtain of mucus between the rim of the shell and the substratum. The mucous curtain is secreted by the mantle, which then withdraws leaving the mucus to dry out. Perforations made experimentally in the mucous curtain are quickly repaired, but permanent obliteration of the curtain causes a seven-fold increase in desiccation rate (Wolcott, 1973).

4.6 Shell Secretion

Shell secretion involves the production of an organic matrix and the deposition of calcium carbonate crystals (section 1.2). By measuring the growth rate of the shell, together with its organic content (1.1 per cent by weight), Paine (1971) estimated that secretion of the organic matrix by *Tegula funebralis* accounts for about 0.8 per cent of its absorbed energy. Palmer (1983b) estimated that the energetic cost of calcification in gastropods is less than 14 per cent, perhaps even less than 3 per cent, of protein synthesis, in which case the metabolic cost of calcification in *T. funebralis* would be less than 0.1 per cent of its absorbed energy.

Secretion of the organic matrix therefore constitutes the major energetic cost of shell secretion and is a significant portion of the energy budget. Not surprisingly, gastropods with a smaller fraction of protein in their shell are able to regenerate broken shell faster than those with more protein (Figure 4.11). The relative expense of the organic matrix may partly explain the repeated evolutionary loss among gastropods of nacre, which contains a greater proportion of organic matrix than other shell materials (Taylor and Layman, 1972). Nacre is a particularly strong shell material (Currey and Taylor, 1974) but evidently in some lineages the energetic cost of its production has outweighed its advantages (Palmer, 1983b).

Energy need not be the only 'cost' of shell secretion. *Nucella lamellosa* occurs as thick- and thin-shelled morphs at similar tidal heights. Thick-shelled morphs have 50 per cent heavier shells than the thin-shelled morphs of similar size, but both secrete shell at the same rate. However, thick-shelled morphs feed and gain body

Figure 4.11: The Rate of Shell Secretion among Different Gastropods Is Inversely Proportional to the Amount of Organic Matrix in the Shell. This has been demonstrated by breaking the lip of the shell and measuring the rate of regeneration. The data are mean regeneration rates for species representing the archaeo-, meso- and neogastropods. After Palmer (1983b)

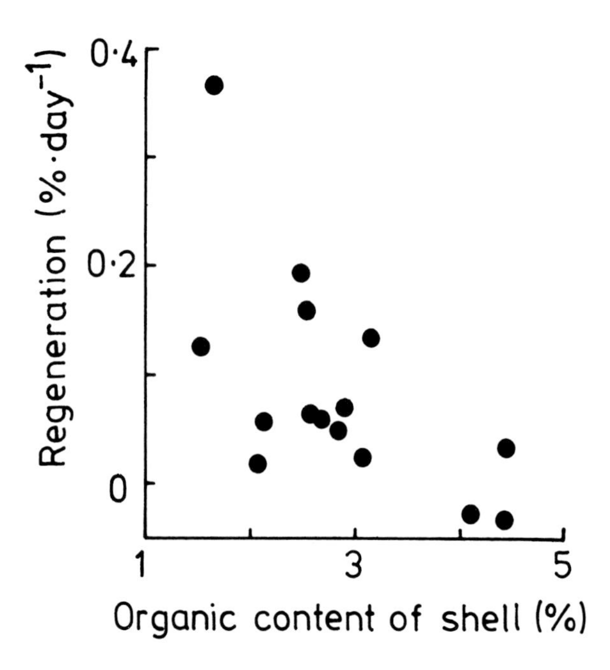

weight slower than thin-shelled ones, probably because thicker shells have less room for body growth. If so, this represents a non-energetic 'cost' of a more robust exoskeleton (Palmer, 1981).

The converse situation, in which body growth determines shell morphology, is exemplified by *Littorina littorea.* As in *Nucella,* faster-growing snails have thinner shells, but they also have less calcium carbonate per unit mass of organic matrix than slower-growing snails derived from the same population (Kemp and Bertness, 1984). The shells of faster growing *L. littorea* are more globose, increasing the internal volume per unit mass of shell. These factors will allow fast growth to continue when calcification is limited to a constant rate.

For any coiled gastropod shell, there is an optimum amount of overlap between successive whorls that minimises the ratio of area of shell material to volume enclosed, but among terrestrial gastropods and probably also among marine ones, the observed amount of overlap always exceeds the theoretical optimum (Heath, 1985). The various shapes of gastropod shells probably result from several, perhaps conflicting selective forces such as resistance to crushing or the need to accommodate a large foot for stronger adhesion to the substratum (section 7.4.3), and this may necessitate the secretion of extra shell material.

5 GROWTH

5.1 Introduction

The balance of absorbed energy left after expenditure on metabolism, excretion and secretion is available either for the production of somatic tissue and nutrient stores or the production of reproductive tissues and gametes. Energy committed to somatic growth cannot simultaneously be used for reproduction, although storage products may be mobilised for reproductive purposes at a later time. Though functionally distinct, somatic growth and reproduction are difficult to separate in practice because measured increments of body weight usually include the biomass of reproductive organs, inextricably mingled with the viscera in most gastropods. Estimates of reproductive energy expenditure in gastropods are therefore normally confined to the biomass of spawned material or released young, whereas estimates of somatic growth include the biomass of the reproductive tracts.

Since metabolism accounts for the greater part of absorbed energy (Table 5.1), the amount of energy available for growth and reproduction (scope for growth) is largely determined by the rates of ingestion and respiration. Both processes are functions of body size, temperature and food supply (Chapters 2 and 4), hence growth and reproduction are likely to be seasonal and dependent on age.

5.2 Relationship between Shell and Tissue Growth

Growth in snails is most easily measured as a linear increment of the shell, converted to somatic growth by using the relationship between body (tissue) weight and shell size. If snails keep the same shape throughout life (isometric growth), then the body weight is directly proportional to the internal volume of the shell and hence to the cube of any of its linear dimensions. This is often approximately true, but changes in size-specific body weight interplay to cause slight allometric deviations from the 'volume law' (Table 5.2). Regressions of body weight on shell size have therefore to be

Table 5.1: Energy Budgets for Gastropods. For comparability, all components are expressed as proportions of ingested energy. These figures are very approximate because they represent population means that obscure the effects on energy partitioning of individual size and reproductive condition. Because of unavoidable imprecision, completely measured energy budgets do not balance exactly. Numbers in parentheses are not measurements, but estimations made by balancing the budget

Species	Ingestion	Defecation	Respiration	Growth	Reproduction	Reference
Patella vulgata	1	(0.55); 0.04*	0.31	0.04	0.06	Wright and Hartnoll (1981)
Fissurella barbadensis	(1)	0.66	0.25	0.08	0.01	Hughes (1971b)
Nerita versicolor	(1)	0.61	0.34	0.04	0.01	Hughes (1971a)
N. peloronta	(1)	0.57	0.37	0.05	0.11	Hughes (1971a)
N. tessellata	(1)	0.60	0.35	0.03	0.01	Hughes (1971a)
Tegula funebralis	1	0.30; 0.07*	0.54	0.04; 0.01†	0.01	Paine (1971)
Polinices alderi	1	Not given	0.21	0.03	0.25	Ansell (1982)
Navanax inermis	1	0.30	0.29	0.09	0.16	Paine (1965)

* = Nitrogenous excretion
† = Shell growth

Table 5.2: Exponent (*b*) Relating Body Weight to Linear Dimension in Gastropods, Defined by the Power Function: Body Weight = *a* (Linear Dimension)^b

Species	Exponent	Reference
Patella cochlear	2.5	Branch (1975)
P. vulgata	3.7	Wright and Hartnoll (1981)
Cellana tramoserica	3.5	Parry (1978)
Fissurella barbadensis	3.3	Hughes (1971b)
Lacuna parva	3.3	Ockelmann and Nielsen (1981)
L. pallidula	2.19	Grahame (1977)
L. vincta	2.84	Grahame (1977)
Nerita tessellata	3.5	Hughes (1971a)
N. versicolor	3.1	Hughes (1971a)
N. peloronta	3.4	Hughes (1971a)
Littorina littorea	3.8	Grahame (1973)
L. littorea	3.4	Hughes and Roberts (1980a)
L. neritoides	2.5	Hughes and Roberts (1980a)
L. nigrolineata	2.7	Hughes and Roberts (1980a)
L. saxatilis male	3.2	Hughes and Roberts (1980a)
L. saxatilis female	2.7	Hughes and Roberts (1980a)
Melagraphia aethiops	3.1-3.7	Zeldis and Boyden (1979)
Polinices duplicatus	2.6	Edwards and Huebner (1977)
Nucella lapillus	3.1	Hughes (1972)
Spiratella retroversa	3.2	Connover and Lalli (1972)
S. helicina	3.5	Connover and Lalli (1972)
Onchidoris bilamellata	2.8	Todd and Doyle (1981)

fitted in each case. Since tissues have a similar specific gravity to water, separate measurements of shell and somatic mass can be made by weighing live snails in water and in air, using a calibration to convert immersed to dry shell weight (Palmer, 1982a).

Estimating somatic growth from shell growth assumes a direct relationship between the two. Over sufficiently long time intervals this is close enough to reality, but shell growth can occur even during starvation when there is no 'scope' for tissue growth (Palmer, 1981) and at other times may occur in rapid spurts uncorrelated with tissue growth. Moreover, changes in biomass, caused, for example, by the accumulation of storage products, may occur without shell growth. For these reasons shell growth could be a misleading index of tissue growth over short time intervals.

5.3 Shell Growth

Linear increments in shell size are the result of new material being secreted by the mantle around the lip of the shell; a process that is interrupted whenever the mantle withdraws from the aperture. Thinly ground sections taken along the direction of growth reveal fine bands in the shells of intertidal gastropods (Figure 5.1A), corresponding to periods of tidal emersion when the mantle margin withdraws from the shell lip. Rough weather, extreme temperatures and starvation are likely to cause inactivity and withdrawal of the mantle. Consequently, prolonged periods unfavourable for shell growth, as in cold winter months, will be represented by closely packed bands in the shell material. Microscopic growth bands therefore record the growth history of the gastropod and would provide very useful research material were it not for the difficulty of taking standard sections exactly along the spiral direction of growth. Externally visible growth bands, demarcating seasonally reduced growth, are sometimes present (Figure 5.1B) and can be used to estimate annual growth increments.

Other methods of measuring shell growth include sequentially measuring the size of tagged individuals or measuring the ratio of the isotopes ^{16}O and ^{18}O in shell material at different distances from the lip. This ratio changes with temperature, so that in seasonal environments the technique can be used to estimate growth rate, as has been done for *Patella tabularis* (Shackleton, 1973).

In simple helical growth (Figure 5.2A), the internal architecture of the shell remains unaltered, except for a general thickening of the wall by secretions of the inner mantle. Other growth forms require restructuring of the internal architecture to accommodate the growing animal, as occurs during the compressed spiral growth of *Conus* spp. (Figure 5.2B). External structural changes involving the formation of ridges, knobs or spines found in many gastropods, especially tropical species, are mostly defensive in function (Chapter 7).

Shell growth may continue at a declining rate throughout life (indeterminate growth), as in *Littorina littorea*, or it may cease at some stage (determinate growth), usually when the snail reaches sexual maturity, as in *Nucella lapillus*. Cessation of growth at the onset of sexual maturity is accompanied by pronounced thickening of the apertural margin, often with tooth-like sculpturing of the inner rim. These features increase the strength of the shell and may defend the aperture against intrusion by predators. Truly determinate growth, however, is rarely the case: *Nucella lapillus* may reinitiate growth in short spurts after reaching sexual maturity.

Figure 5.1: A. Micro-growth Bands in the Shell of *Littorina littorea*. The snail was given a cold shock and the lip of the shell was filed down to produce a notch that would serve as a time marker. The snail was then subjected to 13 tidal emersions before cutting its shell along a plane at right angles to the direction of growth. The cut surface was ground, polished and coated with acetate, which was peeled off the shell and photographed under the microscope. The file notch and stress band caused by the cold shock are clearly visible, and the 13 tidal growth increments are evident as fine bands. Scale bar = 0.1 mm. After Ekaratne and Crisp (1982). B. Annual growth rings in the shell of *Monodonta lineata*. These rings are formed by the retarding or cessation of growth during winter. Shell height 1.8 cm. After Williamson and Kendall (1981)

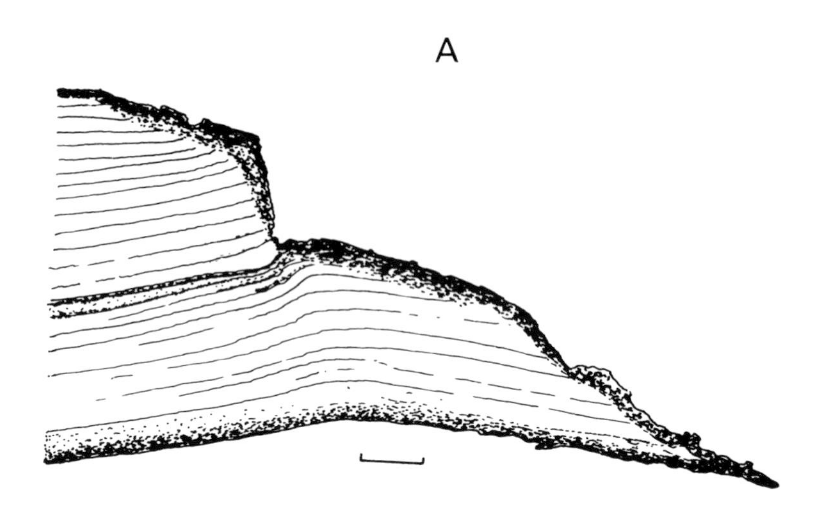

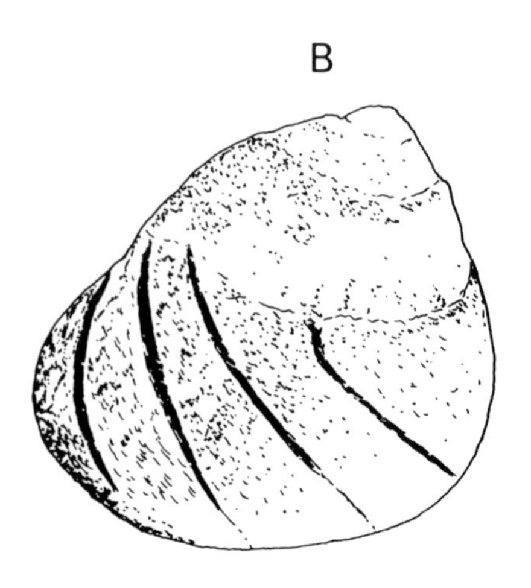

Figure 5.2: A. Helical Growth of Shell Exemplified by *Trochus* sp. Left, X-ray picture revealing the floor of the whorls spiralling round the central pillar (columella). Right, outer surface of shell showing the sutures between whorls. After Meglitsch (1972).
B. Axial Section of *Conus lividus* Showing the Interior Remodelling that Occurs as the Shell Grows. The last whorl of the shell is thick, forming a strong outer wall, but the penultimate whorl is dissolved to a thin septum as the shell grows. Shell material is added to the inside of the spire and to the anterior part of the columella. Thickening of the outer wall, columella and spire enhances protection against crushing predators, and the dissolution of inner whorls greatly expands living space for the snail. IW=Thin inner whorl, OW=thick outer whorl, TC=thickened region of columella, TS=thickened spire. After Kohn *et al.* (1979).

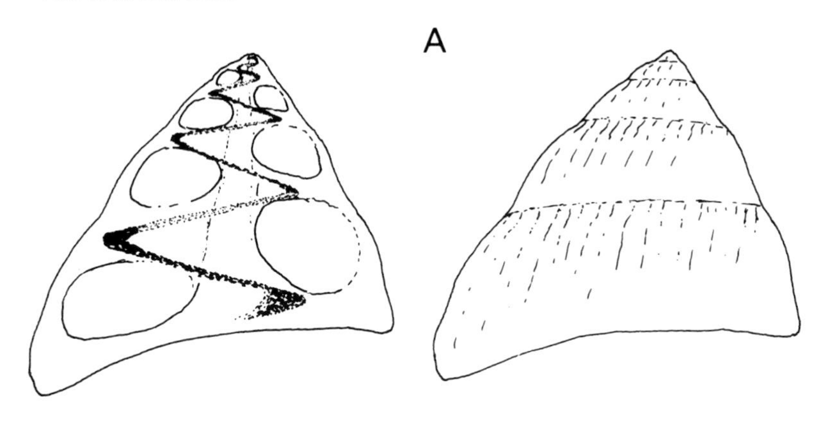

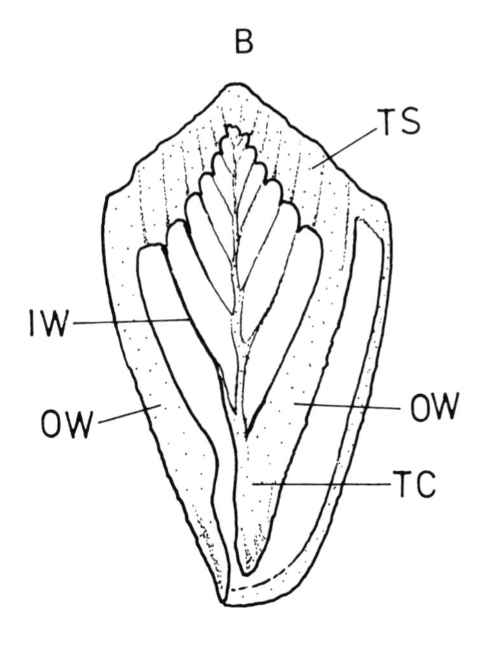

This trend is more pronounced in some species where growth occurs in large steps, each punctuated by thickening of the apertural margin. Previously thickened margins (varices) become ridges as new, thinner shell is extended beyond them in episodes of growth (Figure 5.3). *Cassis tuberosa* is an episodic grower that burrows beneath the sand for protection while secreting new shell, and only emerges to resume foraging once the new margin has been thickened; this requires massive deposition of calcium carbonate and must be about the fastest shell secretion in any mollusc (Linsley and Javidpour, 1980).

Shells are frequently chipped by predators (Vermeij, 1982) or by impact from moving stones (Raffaelli, 1978). Repair is important to restore the protective function of the shell and to accommodate future somatic growth. Accordingly, lost shell material is quickly replaced even during starvation, although under these conditions the rate of secretion is less than when snails are well fed (Palmer, 1983). Forces required to break shells of *Littorina littorea* showing signs of repair indicate that the original strength of the shell is restored (Blundon and Vermeij, 1983).

5.4 Somatic Growth

5.4.1 Growth Curves

Growth in most animals decelerates as they extend beyond a certain size, usually reached early in the juvenile stage. Throughout most of an animal's life, therefore, growth can be described by an asymptotic curve. Such a curve may be derived theoretically by assuming that the rate of energy absorption obeys a 'surface law', being proportional to the square of any linear dimension of the body, and that metabolic rate obeys a 'volume law', being proportional to the cube of the linear dimension (Bertalanffy, 1957): let $W =$ body weight, $L =$ body length and $T =$ time; then, according to the surface and volume laws, absorption $= aL^2$, metabolism $= mL^3$, and growth rate in weight $= aL^2 - mL^3$, but $W = sL^3$ and $\mathrm{d}W/\mathrm{d}L = 3sL^2$. Hence:

$$\text{growth rate in length} = \frac{\mathrm{d}L}{\mathrm{d}T} = \frac{\mathrm{d}W}{\mathrm{d}T} \cdot \frac{\mathrm{d}L}{\mathrm{d}W} = \frac{a}{3s} - \frac{mL}{3s}$$

$$= \frac{m}{3s}\left(\frac{a}{m} - \mathrm{L}\right)$$

Figure 5.3: Episodic Growth in *Phalium granulatum*. After an episode of rapid growth, a thickened rim (varix) is formed around the lip of the shell and there is no more growth until the next episode. At the next episode, new whorls of thinner shell are extended from the old lip, whose thickened rim remains as a prominent ridge (former varix). The specimen has two former varices in addition to the outer thickened lip of the shell, indicating three previous episodes of growth. Top, right lateral; middle, left lateral; bottom, rear view. Shell length 7 cm. SL=Shell lip (varix), VX=former varix

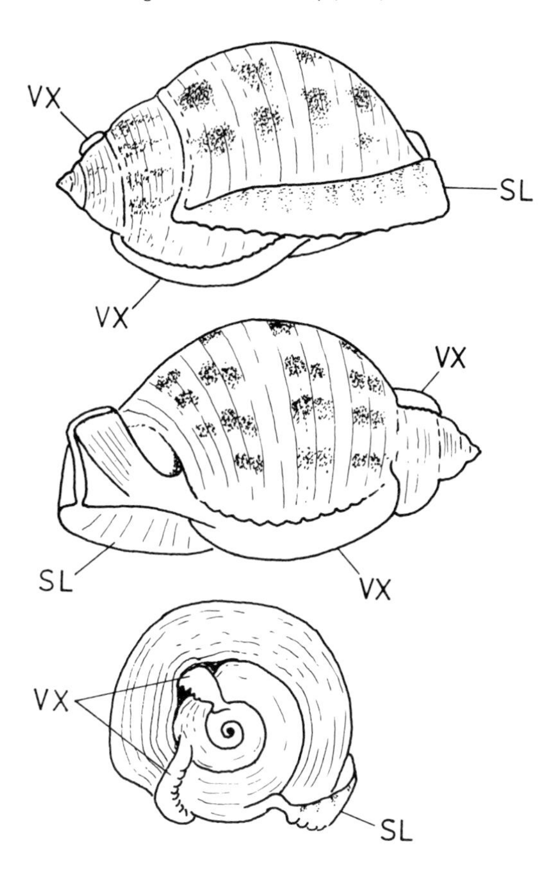

where a, m, s are constants of proportionality, $m/3s$ being the instantaneous length-specific growth rate k, and a/m the asymptotic length L_∞. The integrated form of this equation is:

$$L_{t+1} = L_\infty (1 - e^{-k})_{+e}{}^{-k}. L_t$$

giving a straight line when L_{t+1} is plotted against L_t (Figure 5.4). This is a 'Ford–Walford' plot and is a commonly used method of

Figure 5.4: Annual Shell Growth of *Nucella lapillus* is Shown by a Ford-Walford Plot of Shell Height (L_{t+1}) against Height at the End of the Previous Year (L_t). A regression equation generating the line fitted to the data enables height at any future time to be predicted from any starting height. After Hughes (1972)

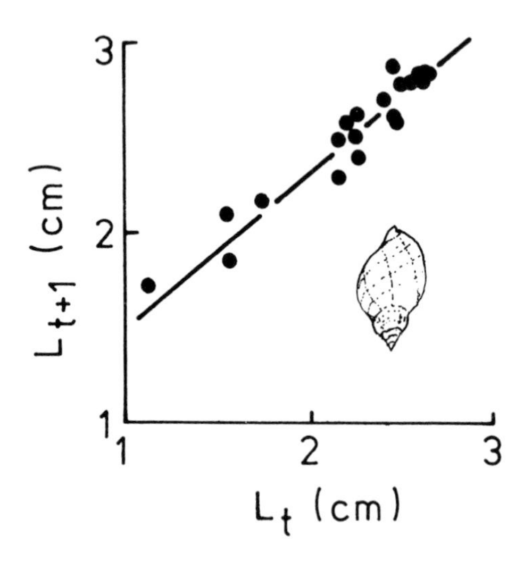

fitting the parameters of the Bertalanffy equation to growth data. Other methods include computerised iterative procedures (Fabens, 1965). It is common practice to apply the asymptotic Bertalanffy equation to the decelerating growth phase of animals, and a good fit is usually obtained, but this is simply because the equation is an asymptotic function and not because it is derived from realistic assumptions. Any departures from the surface or volume law will alter the exponent relating absorption rate or metabolic rate to size, so altering the shape of the growth curve.

Considerable departures from the surface law are found in the ingestion rates of gastropods (section 2.3.1), whereas metabolic rate usually falls somewhere between the expectations of the surface and volume law (section 4.2.1). The net result is a sigmoid growth curve, in which growth accelerates beyond the embryonic stages, reaching a point of inflexion in the early juvenile phase. Although possibly confounded by changes in the shape of the juvenile shell, sigmoid growth can sometimes be detected in the shell if measurements include sufficiently small individuals (Figure 5.5). A sigmoid function that can be fitted to such growth data is the Gompertz equation (Laird *et al.*, 1965), but the goodness of the fit does not imply any biological meaning for the parameters of the equation. Interactions between absorption rate and metabolic rate during ontogeny are probably too complex for representation in a general analytical model, but descriptive models such as the Bertalanffy and Gompertz equations are useful research tools for predicting the size that an animal will reach in a certain time.

A non-linear growth curve implies that the proportion of absorbed energy used for tissue production (production efficiency) changes with body size. In sigmoid growth, production efficiency increases up to the point of inflexion and decreases thereafter, caused largely by different allometric relationships between

Figure 5.5: Evidence of Sigmoid Cumulative Growth in Gastropods. Growth rate of the shell increases at first as the smallest snails become larger, but then declines with further increases in size. The maximum on the growth-rate curve corresponds to the point of inflexion in cumulative growth. Since the increasing phase is confined to the early stages of growth, measurements taken over intervals longer than a few months would obscure the sigmoid relationship and give a straight line with negative slope resembling the Bertalanffy model. A. *Littorina nigrolineata* (data are means) after Hughes (1980). B. *Monodonta lineata* (data are means) after Williamson and Kendall (1981). C. *Cerithium nodosum* (raw data, curve drawn by eye), after Yamaguchi (1977)

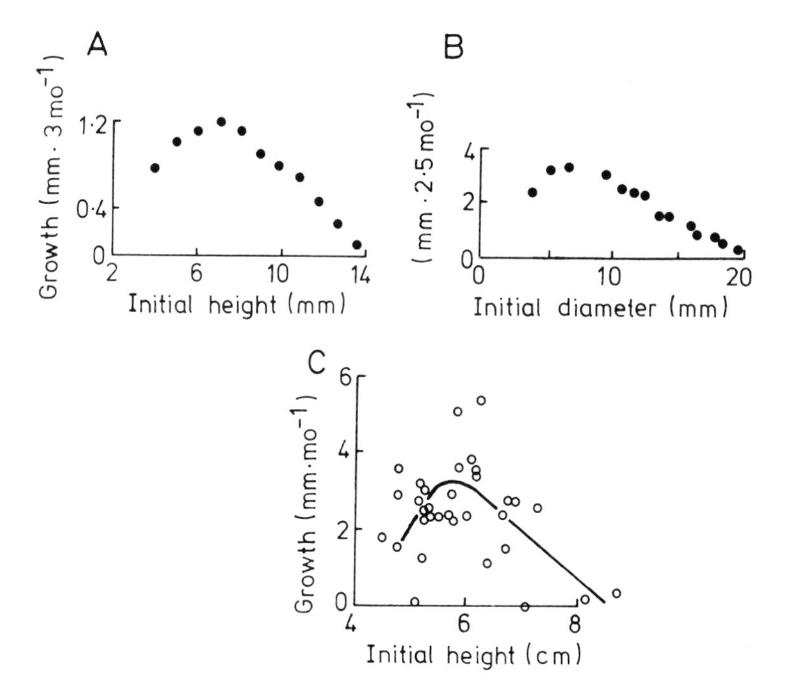

absorption rate, metabolic rate and body size discussed above. Veliger larvae have among the highest production efficiencies recorded for gastropods (Table 5.3), whereas declining production efficiency with increased body size has been recorded in several prosobranchs and nudibranchs (Figure 5.6A).

5.4.2 Growth Efficiency

After sexual maturation, the balance between absorbed and metabolised energy is divided between somatic growth and reproduction, allocation depending on age, life history and environmental factors (Chapter 6). Growth efficiency is strictly defined as the proportion of ingested energy (gross efficiency) or absorbed energy (net efficiency) used for somatic growth, but when, as is often the case, growth rates have been calculated from increments of total body weight, published 'growth efficiencies' are actually production efficiencies.

Up to sexual maturity, production and growth are synonymous, but thereafter an increasing fraction of production is devoted to reproduction and so true growth efficiency declines with age faster than total production efficiency (Figure 5.6B). When somatic growth and gonadal development coincide, the reproductive claim on absorbed energy will depress somatic growth (Figure 5.7), causing seasonal changes in growth efficiency. Single published figures for the production efficiencies of species are therefore

Table 5.3: Production Efficiencies of Gastropods, Defined as the Proportion of Assimilated Energy Used for Growth and Reproduction

Species	Production efficiency (%)	Reference
Patella vulgata	25	Wright and Hartnoll (1981)
Fissurella barbadensis	27	Hughes (1971b)
Nerita tessellata	12	Hughes (1971a)
N. versicolor	13	Hughes (1971a)
N. peloronta	16	Hughes (1971a)
Littorina obtusata	40	Wright and Hartnoll (1981)
Tegula funebralis	13	Paine (1971)
Nucella lapillus	26	Hughes (1972)
Navanax inermis	49	Paine (1965)
Archidoris pseudoargus	63	Carefoot (1967b)
Dendronotus frondosus	28	Carefoot (1967b)
Aplysia punctata	38	Carefoot (1967b)
Veligers of:		
Littorina littorea	60	Jørgensen (1952)
Nassarius reticulatus	60	Jørgensen (1952)

Figure 5.6: A. Gross Growth Efficiency (Somatic Growth/Ingestion) Decreases Exponentially with Increasing Age in *Polinices duplicatus*. The semi-logarithmic plot has transformed the exponential curve into a straight line. After Edwards and Huebner (1977). B. Net Growth Efficiency (P_g/A) Decreases Exponentially with Increasing Age in *Patella longicosta*. Net production efficiency (P/A) decreases more gradually because it includes reproductive effort (P_r/A), which increases with age until the limpet begins to senesce. After Branch (1981)

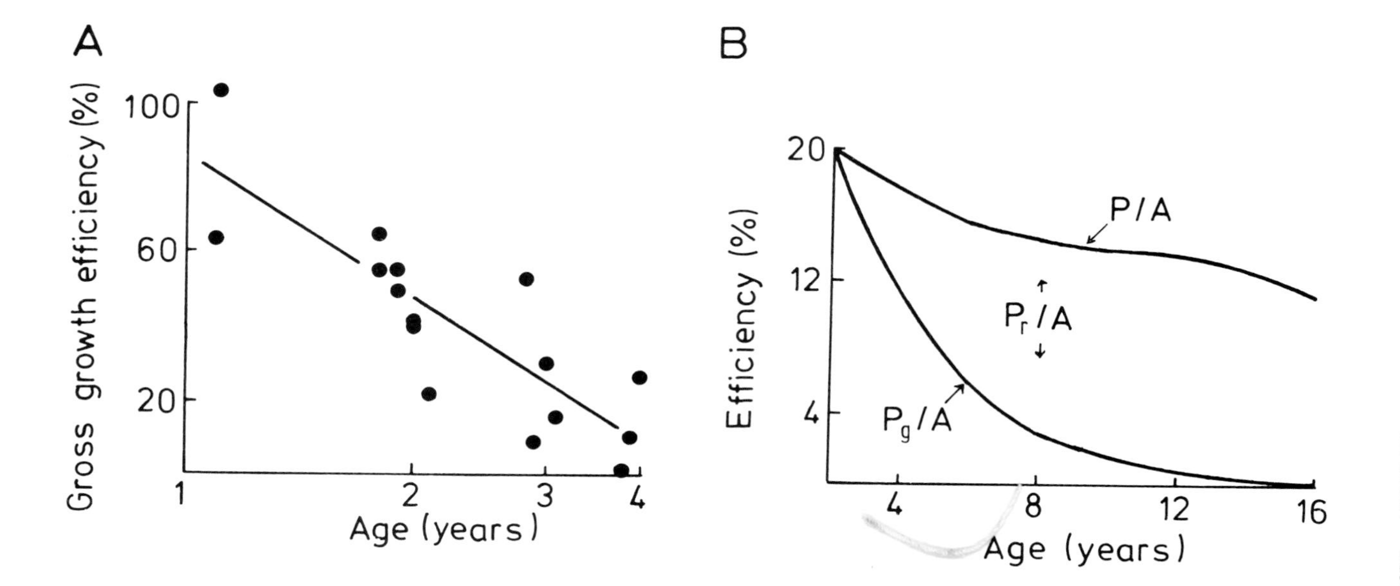

crude averages that obscure seasonal variation and are probably biased downwards towards adult mean values in many cases. Most single-figure estimations of production efficiency lie between 10 and 60 per cent, longer-lived, slower-growing species such as *Tegula funebralis* and *Nucella lapillus* having lower production efficiencies than shorter-lived, faster-growing ones such as *Aplysia punctata* and *Navanax inermis* (Table 5.3). Correspondingly, the mean productivity per unit body weight declines as longevity increases and is allometrically related to lifespan by an exponent of about 0.7 (Figure 5.8).

Production depends on the food supply exceeding the metabolic demands of the animal and is therefore likely to be affected by temperature, food quality and food abundance. At higher temperatures *Limapontia capitata* must feed faster to achieve the same growth rate as at lower temperatures because metabolism is faster and claims more of the absorbed energy (Figure 5.9). Among individuals of *Clione limacina* feeding in the laboratory on *Spiratella retroversa*, those individuals ingesting more food per day grow faster (Figure 5.11). *Aplysia punctata* produces biomass faster and with greater efficiency when feeding on preferred algal species,

Figure 5.7: Net Growth Efficiency of *Polinices alderi* Declines with Increasing Body Size and is Generally Lower when the Snail Enters its Reproductive Phase (Dotted Line) because Less Absorbed Energy is then Available for Growth. P_g=Somatic productivity, P_r=reproductive productivity. The curves represent snails kept at 20°C. After Ansell (1982)

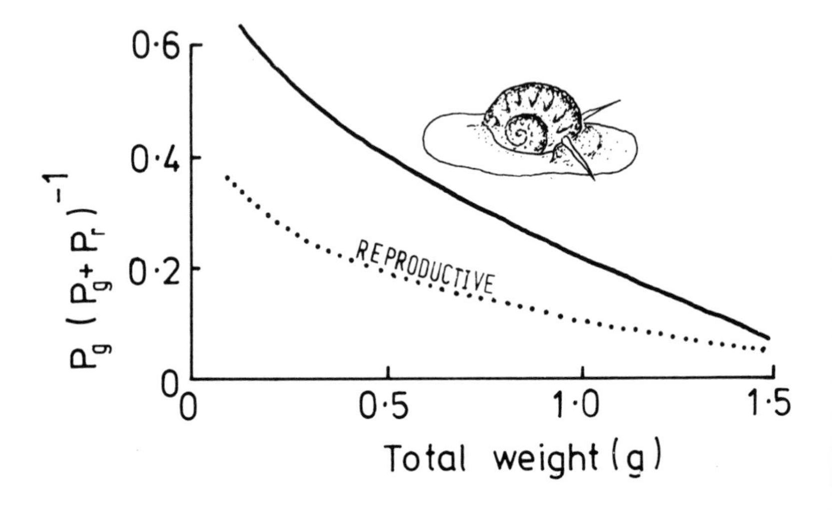

Figure 5.8: Production per Unit Biomass (P/B) Is Inversely Proportional to Average Life Span. Closed circles = species of limpet (after Branch, 1981), open circles = other prosobranchs (after Robertson, 1979). The data lie reasonably close to Robertson's (1979) regression based on various benthic invertebrates, which has a slope of −0.73

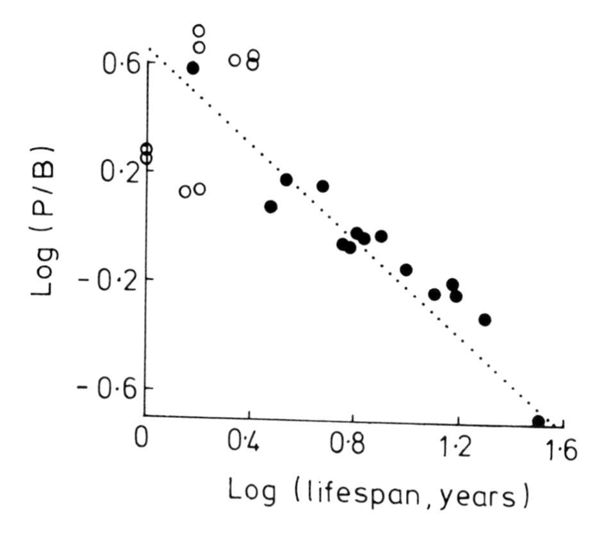

Figure 5.9: Increasing Temperature Decreases the Gross Growth Efficiency of *Limapontia capitata*, so that to Achieve the Same Growth Rate the Slugs Must Consume More Food at Higher Temperatures. Regressions were fitted to data on the number of cells of *Cladophora* (green filamentous alga) eaten per slug and the volumetric increase of the slug over 8 days. After Jensen (1975)

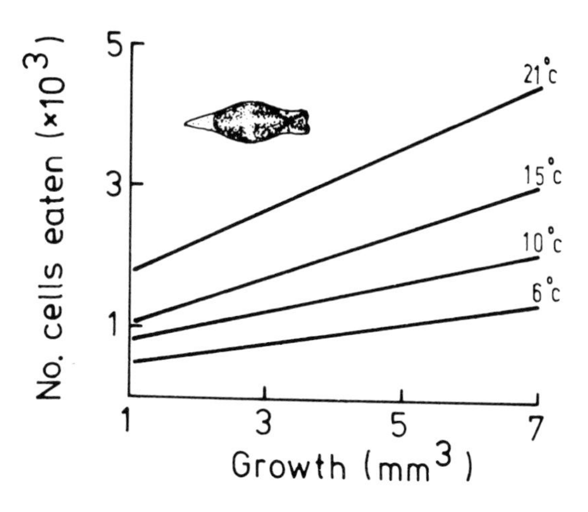

Figure 5.10: A. *Aplysia punctata* Grows Faster on a Diet of its Preferred Food, *Plocamium coccineum*, than on the Successively Less Preferred Algae *Ulva lactuca* and *Cryptopleura ramosa*. Each line represents the mean of 6 slugs. After Carefoot (1967a). B. *Olea hansineensis* Sustains Better Growth on a Diet of its Preferred Food, the Eggs of *Haminoea virescens*, than on a Diet of Less Preferred *Melanochlamys diomeda* Eggs. The diet of *H. virescens* eggs was preceded by a week of feeding on the less preferred *Archidoris montereyensis* eggs and this seems to have slightly depressed growth during the first month. Each line represents the mean of 20 slugs. After Chia and Skeel (1973)

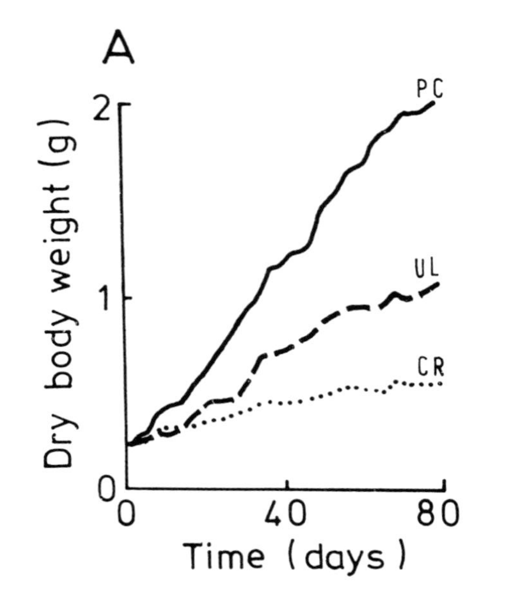

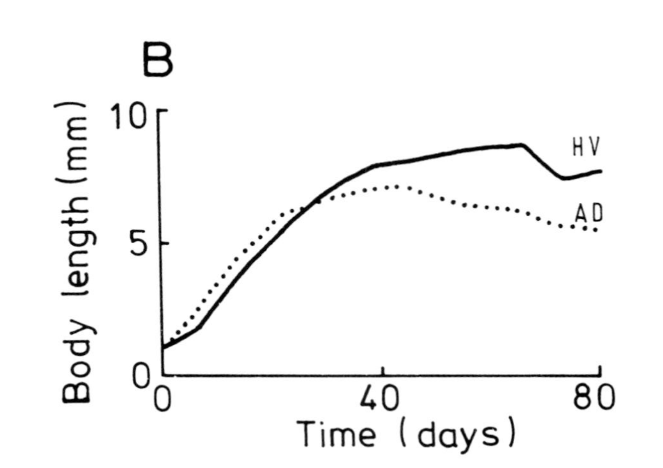

Figure 5.11: Growth Rate is Linearly Related to Feeding Rate in *Clione limacina* Preying upon *Spiratella retroversa*. Growth rate was calculated as $1/t.\log_e (W_2/W_1)$, where W_1 =initial weight (mg), W_2=final weight, and t=growth period. Feeding rate (at 20°C) was calculated as $1/t.\log_e [(W_1 + W_S)/W_1]$, where W_1=initial weight of *Clione*, W_S=weight of *Spiratella* eaten, t=feeding period. After Connover and Lalli (1972)

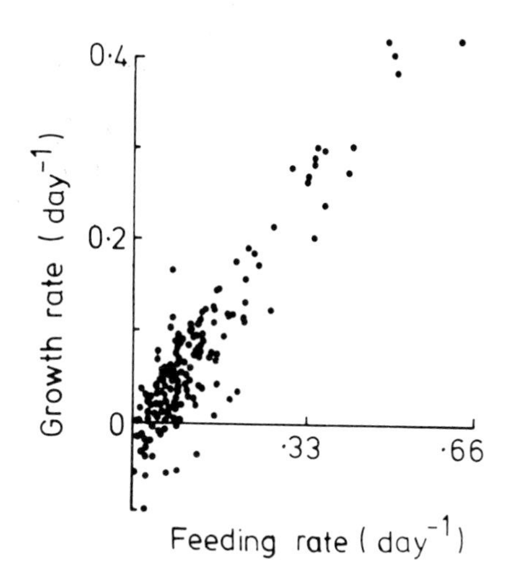

Figure 5.12: The Growth Rate of *Nerita atramentosa* Is Correlated with the Algal Biomass on Intertidal Rock Surfaces. At higher levels on the shore there is generally less algal biomass, and snails living there grow slower than those lower on the shore. On the figure, growth rate of *N. atramentosa* is expressed as a dimensionless number derived from the exponential growth equation: growth rate = $\log_e(L_t/L_0)/t$. Algal biomass is expressed as the chlorophyll content of scrapings taken from the rock surface. Numbers denote shore levels from 1 = lowest to 4=highest. After Underwood (1984)

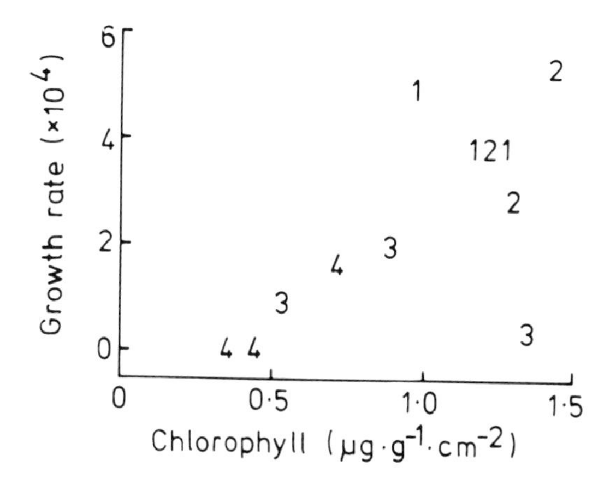

which are absorbed more efficiently, than when feeding on less preferred species (Figure 5.10A). *Olea hansineensis* grows larger and spawns more eggs when raised on its preferred food, the eggs of *Haminoea virescens*, than when given the eggs of *Melanochlamys diomeda* (Figure 5.10B). Similarly, *Onchidoris aspersa* grows about six times faster when maintained on its preferred prey *Electra pilosa* than when maintained on *Hippothoa hyalina* (Smith and Sebens, 1983).

Limited food supply probably retards growth of individuals in many natural populations of gastropods. On certain Australian shores *Nerita atramentosa* and *Bembicium nanum* grow slower high on the shore than lower down, correlated with the amount of micro-algal growth on the rock (Figure 5.12). In other localities individuals of *Notoacmea petterdi* grow faster higher on the shore than at lower levels where their greater density causes intensive competition for food (Creese, 1981).

6 REPRODUCTION

6.1 Introduction

By definition, genetic fitness is maximised when an organism contributes the greatest possible number of recruits to the next reproductive generation, achieved both by increasing the number of young produced in the parent's life and by increasing the survivorship of the young. But this is not necessarily the same as maximising the number or survivorship of the young produced. Increased expenditure on reproduction leaves less energy for growth that may enhance parental survival and the chances of further reproduction, and increased energetic investment per offspring may elevate juvenile survival but depress parental fecundity (number of eggs per clutch).

Maximising fitness, a corollary of evolution, is therefore an optimisation process balancing the proportional expenditure on reproduction (reproductive effort) and the parental commitment per offspring (parental investment) in combinations that produce the highest average numbers of reproductive recruits. Different environmental circumstances will tend to invoke different optimal solutions, but inherent biological constraints characteristic of individual taxa may impose alternative solutions to similar selection pressures.

6.2 Reproductive Effort

6.2.1 Reproductive Value

Greater reproductive effort increases the number of young in a clutch, whereas greater parental investment per offspring decreases it because the reproductive resources are divided among fewer young. Given a certain level of parental investment, clutch size is directly proportional to reproductive effort and is likely to be adjusted during an organism's life to yield the maximum total number of young produced. This principle is modelled in Fisher's (1930) concept of reproductive value, the parental expectation of young. Fecundity usually changes with age and in gastropods this is often associated with increased gonadal volume as the animal grows (Figure 6.1A,B). Survivorship also changes, especially during the juvenile and senescent phases of life (Figure 6.1C). Thus the age-specific reproductive value (expressed as a mother's expectation of future daughters, since only females in a sexual population can reproduce) is defined as:

$$-V_{x'} = \frac{1}{l_{x'}} \sum_{x=x'}^{n} l_x m_x$$

where x' = present age, n = maximum reproductive age, l_x = proportional survival to age x, m_x = fecundity at age x.

Fecundity and survivorship interplay during the various stages of life, the expected total reproductive output being represented by the area under the curve obtained by plotting reproductive value against age. The shape of the curve is determined by the life history of the animal. In short-lived species, such as *Aplysia punctata* and *Lacuna* spp., which have only one reproductive phase before death (semelparity), reproductive value rises steeply to a maximum and then drops quickly to zero (Figure 6.2A). In longer-lived species with indeterminate growth, such as *Littorina saxatilis*, which normally experience several reproductive seasons (iteroparity), reproductive value rises more gradually and may fluctuate between seasons owing to the interaction between continued risk of mortality and reproductive inactivity (Figure 6.2B). The important point is that overall reproductive value continues to rise with age until the advent of senescence. It is therefore worth while for these snails to increase their reproductive effort as they get older.

Conversely, earlier in life it may be better to invest resources in the soma, which has potential for continued growth, increasing its fecundity and providing further opportunities to reproduce. In snails with determinate growth, fecundity does not increase much during adult life, and reproductive value would be expected to decline slowly as life expectancy gradually decreases with increasing age after sexual maturity.

6.2.2 *Age and Reproductive Effort*

Curves relating reproductive effort to age therefore vary in shape according to schedules of growth and survivorship. *Littorina saxatilis* grows relatively quickly towards its asymptotic size and approaches maximal reproductive effort when about 5 years old,

Figure 6.1: A. Clutch Size (Number of Eggs Spawned per Season) of *Littorina neritoides* Is Linearly Related to Body Size after Logarithmic Transformation of Both Variables. The dotted line has a slope of 3, indicating that clutch size is proportional approximately to the cube of the shell height and hence to the volume of the snail. After Hughes and Roberts (1980b). B. Gonadal Weight of *Fissurella barbadensis* Is Linearly Related to Shell Length when the Variables are Plotted on Logarithmic Axes. A regression fitted to the data has a slope of 3.2, indicating a direct relationship between gonadal weight and body volume. Male (open circles) have similar size-specific gonadal weights to females (closed circles). After Hughes (1971b). C. Survivorship Curve for a Cohort (Batch of Animals of Similar Age) of *Littorina nigrolineata*. On the semi-logarithmic plot, a constant slope represents a constant instantaneous rate of mortality. Mortality is greater among young snails, levelling to a constant value throughout adult life. *L. nigrolineata* seldom survive to the age of senescence, where the curve would become steeper. After Hughes and Roberts (1981)

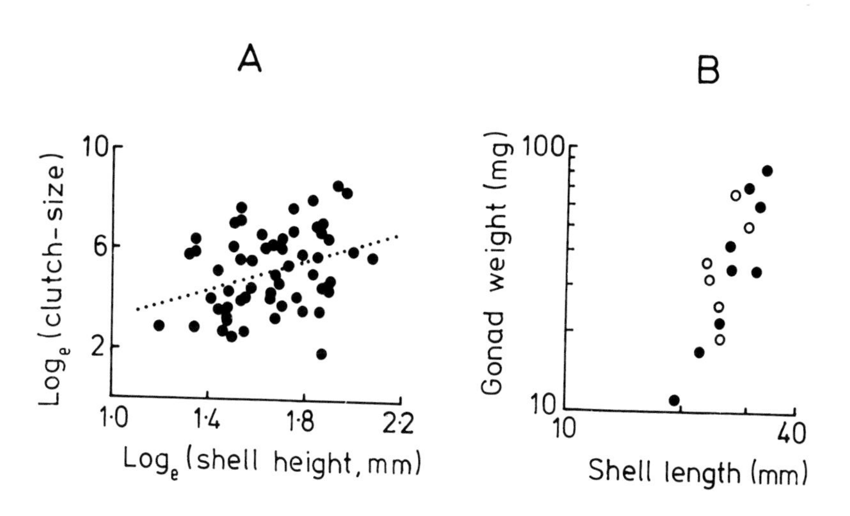

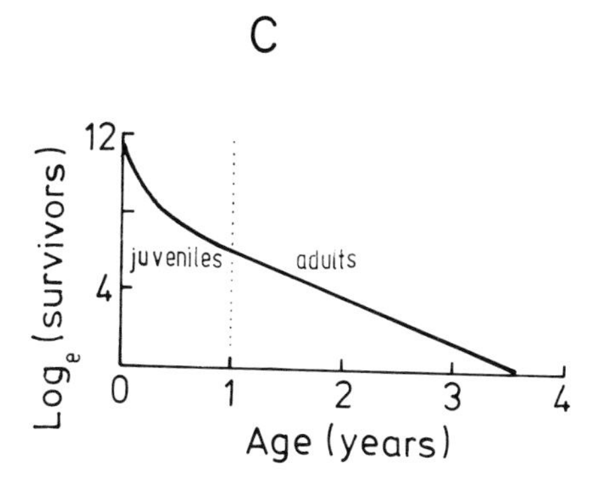

Figure 6.2: Reproductive Value (Expectation of Future Daughters) Is Low in Young Animals because the Probability of death before Reaching Reproductive Maturity Is High. This probability decreases as reproductive maturity is approached and so the reproductive value curve rises with age. A. Semelparous species die after reproducing, so the reproductive value curve drops to zero at the age of reproduction. The curve shown is a theoretical one. B. Iteroparous species live beyond the age of first reproduction and, as in this case of *Littorina saxatilis*, if fecundity increases with age, the reproductive-value curve continues to rise. Eventually, when the probability of death increases but fecundity no longer increases, the reproductive-value curve descends towards zero. Notches in this curve of *L. saxatilis* correspond to seasons when reproduction ceases but the risk of mortality continues unabated. After Hughes and Roberts (1981)

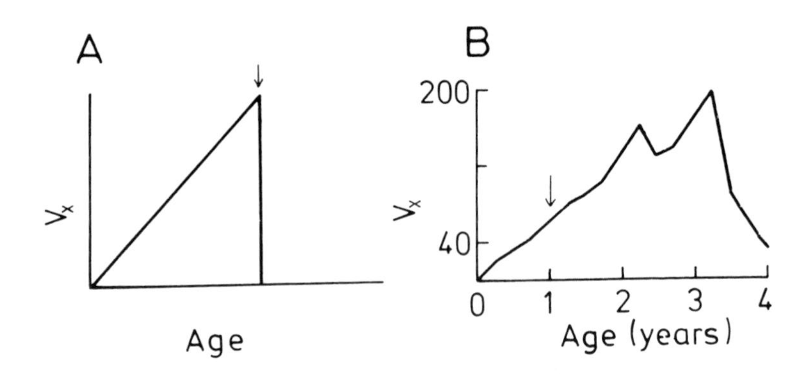

whereas *L. neritoides* grows slower, has a longer mean lifespan and approaches maximal reproductive effort at about 12 years of age (Figure 6.3A). Of course, individuals may not survive to the age of maximum reproductive potential: in the populations studied, only 0.9 per cent of *L. saxatilis* survived for 5 years and only 0.001 per cent of *L. neritoides* survived for 12 years (Hughes and Roberts, 1981).

When the time scale is standardised for growth rate, *L. neritoides* is seen to increase its reproductive effort faster than *L. saxatilis* during equivalent phases of growth (Figure 6.3B). Moreover, when the time scale is standardised for generation time, a demographic parameter that takes into account survivorship as well as growth rate, *L. neritoides* and *L. saxatalis* appear to have similar reproductive-effort curves (Figure 6.3C).

These comparisons suggest that reproductive effort in littorinids is adjusted to the conditions for growth and survival in particular habitats. *L. neritoides* lives high on exposed shores where algal growth is sparse and weather conditions may prevent feeding for

Figure 6.3: Reproductive Effort (Proportion of Total Production Committed to Reproduction) Increases with Age in the Iteroparous Snails *Littorina saxatilis* (Continuous Line) and *L. neritoides* (Dotted Line). A. At similar ages measured in years, the shorter-lived *L. saxatilis* exerts higher RE than *L. saxatilis*. B. At similar 'ages' measured as equivalent stages in growth, *L. neritoides* exerts higher RE than *L. saxatilis*. C. At similar 'ages' measured in units of the mean generation time for each population, *L. saxatilis* and *L. neritoides* exert similar RE. After Hughes and Roberts (1980a)

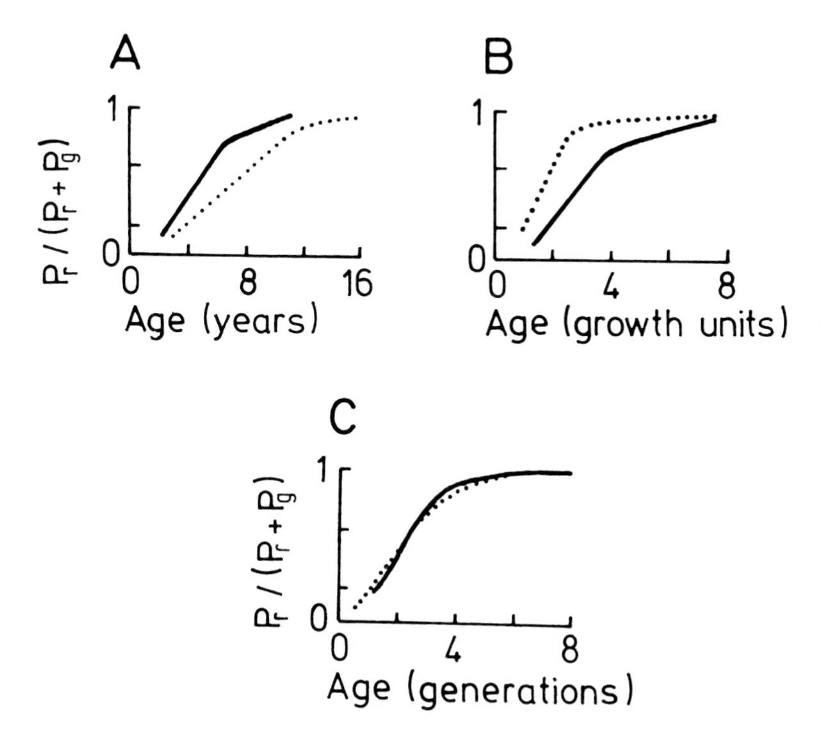

extensive periods, but where protective crevices reduce risks of predation or physical damage. The ecotype of *L. saxatilis* represented in Figure 6.3 lives on boulder shores where algal growth is better and feeding is less extensively interrupted by the weather, but where crab predation and storms cause a higher level of mortality (Hughes and Roberts, 1981).

As expected, species with limited increments in fecundity between spawning seasons exert relatively constant reproductive effort from year to year. This is exemplified by muricids, *Conus* spp. and various limpets (Fletcher, 1984).

6.2.3 Survivorship and Reproductive Effort

The effect of survivorship on reproductive effort has received considerable theoretical treatment (reviewed in Stearns, 1976; Calow, 1983). Members of populations repeatedly depressed to low densities are expected to have higher reproductive efforts than members of more stable populations. This may be true of some species: *Littorina littorea* and *L. neritoides* both liberate numerous planktonic eggs, but *L. littorea* lives lower on the shore where risks of heavy mortality in winter storms are greater and has a higher level of reproductive effort (Hughes and Roberts, 1980a); South African species of *Patella* that migrate upshore as they grow suffer greater mortality and exert greater reproductive effort than territorial species lower on the shore (Branch, 1981); *Hydrobia neglecta* and *Potamopyrgus jenkinsi* in Danish brackish-water habitats have less stable populations and higher levels of reproductive effort than *H. ventrosa* and *H. ulvae* in comparable habitats (Lassen, 1979); reproductive effort is correlated with rates of mortality caused by environmental hazards, starvation or predation among certain Australian limpets (Figure 6.4).

In other cases there appears to be no relationship between reproductive effort and adult survivorship. *Littorina saxatilis* is a polymorphic species living on a wide variety of shores, but reproductive effort is, if anything, lower in populations inhabiting

Figure 6.4: During the Reproductive Season, *Cellana tramoserica* Drains its Energy Reserves to Such an Extent that it May Die of Starvation. There is a positive correlation between reproductive effort (measured as the percentage of absorbed energy used in reproduction) and the mortality rate caused by extrinsic factors such as starvation. After Parry cited in Branch (1981)

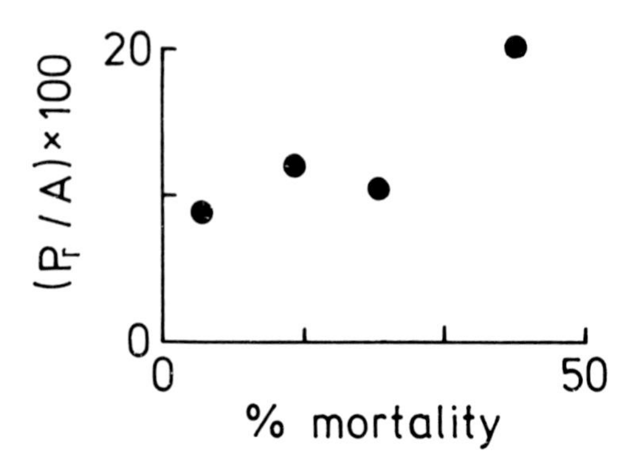

environmentally unstable boulder shores than in those inhabiting more stable saltmarshes (Hughes and Roberts, 1980a).

Sometimes trends in reproductive effort may be caused by environmental constraints on the organism rather than by differential mortality rates among juveniles and adults (Calow, 1983). On a gently sloping rocky beach and on an adjacent vertical cliff in North Wales, populations of *L. saxatilis* have different, genetically determined life-history characteristics (Begon and Mortimer, 1981). Wave impact is often severe in both habitats. The cliff population is confined to protective crevices that will not accommodate large snails. In this population, the individuals would gain nothing by sacrificing reproductive effort for enhanced growth, and instead selection has favoured smaller size, earlier sexual maturity and greater reproductive effort. On the rocky beach, crushing by wave-tossed boulders is a frequent risk that is lessened by growing larger with a concomitantly stronger shell. Energy devoted to large size and a thicker shell must detract to some extent from reproduction so selection has favoured larger size, later sexual maturity and lower reproductive effort in this population.

Apparently closer to the predictions of conventional theory, the small prosobranch *Alvania pelagica*, which despite its specific name is a benthic deposit-feeder, has a larger adult size, longer life and greater reproductive effort in deeper water off the eastern North American continental slope, where physical conditions are less vigorous, than in shallow water (Rex, 1979).

There is therefore some evidence among gastropods that, other things being equal, reproductive effort increases with decreasing adult, relative to juvenile, survivorship in accordance with conventional theoretical predictions. Other things, however, are seldom equal, and this underlying trend is often overshadowed by the effect of different factors such as environmental or phylogenetic constraints.

There may also be an underlying trend for reproductive effort to decline as body size increases. Allometric relationships, largely adduced from mammalian data, suggest that reproductive effort is proportional to (body weight)$^{-0.2}$, absolute reproductive expenditure being relatively constant at a given body size among different species (Peters, 1983). The validity of this conclusion has not been tested for gastropods.

6.2.4 Semelparity and Iteroparity

Survivorship often differs markedly between juvenile and adult phases of the life cycle and the nature of this difference may influence reproductive effort. If juveniles are inherently likely to survive better than adults, then life histories are expected to tend towards semelparity (one reproductive phase, followed by death) and higher reproductive effort; conversely, if adults are likely to survive better than juveniles, the expected tendency is towards iteroparity (repeated reproductive phases during the life cycle) and lower reproductive effort.

Evidence that this theory holds true for gastropods is not unequivocal, however. Larval gastropods are inherently much more vulnerable than adults, and, as expected, iteroparity predominates among prosobranchs that produce tiny planktonic veligers. But semelparity is typical of nudibranchs, not only among those with direct development, but also among the many species with planktotrophic larvae (Todd, 1981). Many of these species exploit ephemeral or pulsed food resources: most nudibranchs feed specifically on sponges, hydroids, bryozoans or tunicates that occur in patches, which have a limited capacity to sustain predation and can only be reached by larval dispersal, or that are seasonal in abundance. The ephemeral and patchy nature of the food may curtail adult lifespan and so enforce semelparity.

The distinction between semelparity and iteroparity is not always obvious. Some species may reproduce continually from year to year, albeit at a fluctuating rate, and therefore exhibit both a single reproductive bout, as required by the definition of semelparity, and reproduction over more than one season, as required by the definition of iteroparity. For example, in Californian populations, the limpet *Notoacmea scutum* is reproductively active throughout the year, but with peak spawning in late spring and summer (Phillips, 1981). Species such as these, however, are best regarded as being iteroparous because there is no evidence that their reproductive bout is followed by physiological deterioration and death, as would be the case with true semelparity.

6.2.5 Reproductive Expenditure per Unit Body Mass

In addition to sequestering absorbed energy, reproduction may also use resources stored in somatic tissues. Generally there is a seasonal accumulation of nutrients in the visceral mass which are eventually used in gametogenesis and oviposition. But since gastropods generally lack discrete nutrient depots (vesicular connective tissue cells, perhaps acting as nutrient depots, have been described in pulmonates (Sminia, 1972)), storage occurs primarily through cellular proliferation (Stickle, 1975). Muricids produce substantial proteinaceous egg capsules, and seasonal fluctuations of nutrient resources are most apparent in the protein fraction, whereas in *Littorina littorea*, producing small egg capsules, and *Patella vulgata*, producing no capsules, seasonal fluctuations are more pronounced in the carbohydrate and lipid fractions (Figure 6.5).

Beyond a certain level, use of somatic resources for reproduction will depress survivorship, since the body has less in reserve for buffering the effects of environmental stress. At mid-shore

Figure 6.5: A. *Nucella lamellosa* Mainly Uses Protein as a Source of Energy for Overwintering and Reproduction. Levels of carbohydrate and lipid show little seasonal change. After Stickle (1975). B. *Littorina littorea* Uses Carbohydrate and Lipid Much More than Protein as a Source of Energy. Levels of carbohydrate and lipid build up to a peak before winter and the spring reproductive period. After Williams (1970). C. *Patella vulgata* also Mainly Uses Carbohydrate and Lipid as Energy Reserves. After Blackmore (1969)

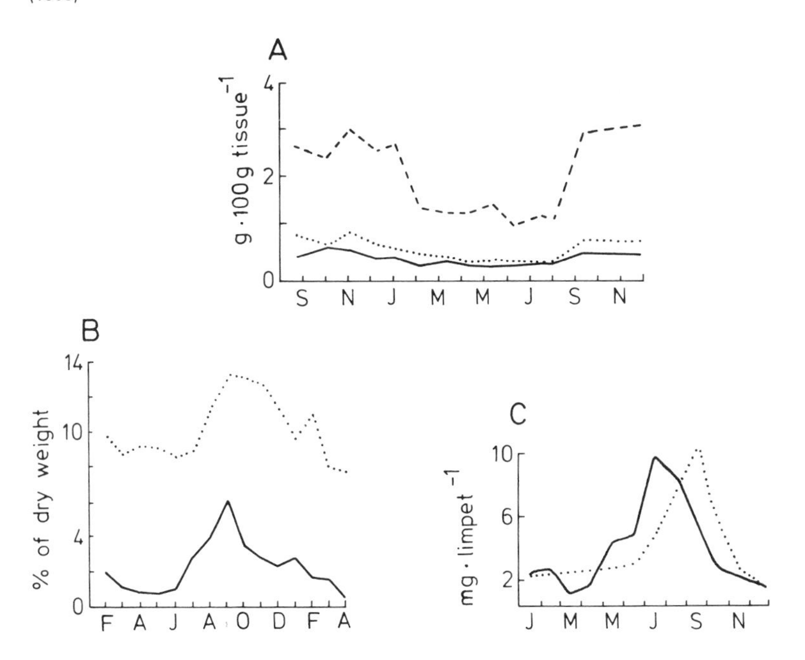

levels, *Cellana tramoserica* sometimes starves to death when algae die back in the summer. Reproduction before the period of starvation depletes resources that might otherwise have been used to withstand starvation (Parry, quoted in Branch, 1981).

Extreme reproductive depletion of somatic resources will itself be lethal and this occurs in semelparous species, which are therefore expected to have higher reproductive expenditures per unit body mass than iteroparous species. Such is generally the case, but there is no clear separation between the two reproductive categories. Some iteroparous species, for example *Nucella lamellosa* and *N. lapillus*, which are long-lived but grow very little as adults, seasonally accumulate large nutrient reserves that are committed almost entirely to reproduction without eroding the basic somatic resources (Figure 6.5A), yet weight-specific reproductive expenditure is high, approaching that of semelparous species (Figure 6.6). It must be borne in mind, however, that energy inputs and metabolic outputs are non-linear and are different functions of body mass, making weight-specific reproductive expenditure a somewhat unreliable comparative index (Calow, 1983).

Figure 6.6: Reproductive Effort Tends to Be Greater among Semelparous than among Iteroparous Snails, but there is No Sharp Distinction between the Two. To standardise for size among species, reproductive effort is expressed as the weight of eggs or young released per year per unit body weight. To standardise for growth rate, age is expressed in units of equivalent stages on the growth curve. From top to bottom, the curves represent *Lacuna vincta* (semelparous), *L. pallidula* (semelparous), *Nucella lapillus* (iteroparous), *Fissurella barbadensis* (iteroparous), and *Nerita versicolor* (iteroparous). After Hughes and Roberts (1980a)

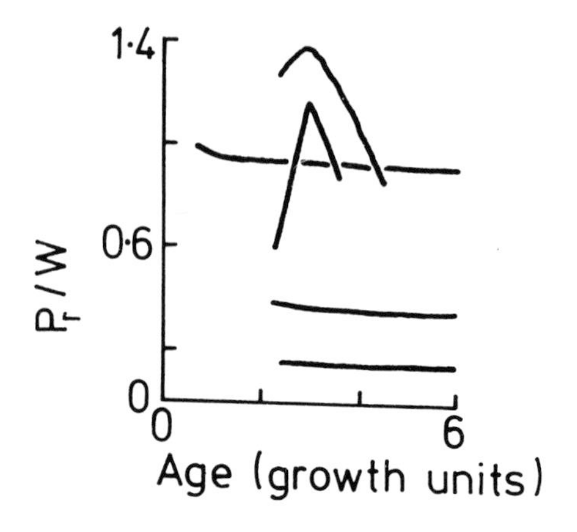

Because continued survival is important in iteroparous species, they may be expected to adjust weight-specific reproductive expenditure according to food availability. Spawning rate in *Nucella lamellosa* thus fluctuates in parallel with the density of barnacle prey (Figure 6.7). The spawn produced per unit body mass by *Cellana tramoserica* is inversely correlated with the availability of algal food (Fletcher, 1984). In sublittoral populations, where density is low and algal growth unlimited by desiccation, limpets spawn the equivalent of 101 per cent of their body weight per season. At high shore levels, where density is also low but where algal growth is limited by desiccation, limpets spawn 51 per cent of their body weight and only 20 per cent at mid-shore levels where density, hence competition for food, is high.

Such a response, however, is appropriate only if adult survivorship remains high. If shortage of food significantly reduces survivorship, then it may be selectively advantageous to increase reproductive expenditure at the expense of the soma, since the

Figure 6.7: In Years when Prey (Barnacles) Are More Abundant, *Nucella lamellosa* Grows Faster and, because of its Increased Size, Lays More Eggs. Fecundity is expressed as the number of eggs laid by a population of *N. lamellosa*. Both fecundity and abundance of barnacles are in arbitrary units. From Spight and Emlen (1976)

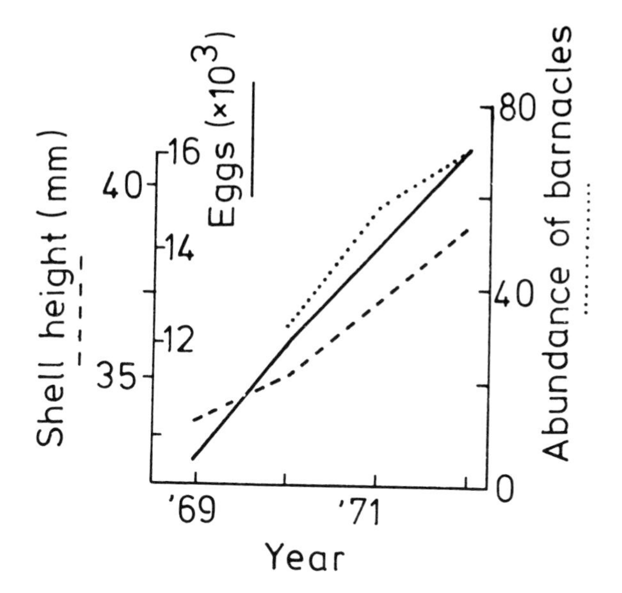

chances of surviving to another season are, in any case, small. On South Australian sand beaches, *Nassarius pauperatus* in populations with less food lay more eggs than those in populations with more food (McKillup and Butler, 1979).

6.3 Male and Female Reproductive Investment

6.3.1 Sexual Differences

Most prosobranchs have separate sexes (gonochoristic) with little or no morphological distinction between males and females except for the reproductive organs. Males may mature at a smaller size and not grow as big as females (Figure 6.8), but the difference is usually slight except among sequential hermaphrodites (Figure 6.10). In some species, such as *Olivella biplicata* (Edwards, 1969), males are larger than females.

Energetic expenditure on gametes by large male *Littorina littorea* is about two-thirds that of similarly sized females, the difference narrowing towards smaller sizes (Figure 6.9). Data for other species are sparse, but large sexual differences in reproductive effort are theoretically not expected. Each sex is a limiting commodity for the other, and intra-sexual competition will ensure that both males and females exert the maximum reproductive effort compatible with other demands on their resources. Therefore, if males and females are morphologically and ecologically similar, they will exert similar reproductive effort. Concordantly with this prediction, testes reach about 40 per cent of somatic weight in large male *Patella vulgata* and ovaries of large females reach about 37 per cent (Wright and Hartnoll, 1981).

6.3.2 Sex Ratio

Not only reproductive effort but also the frequencies of males and females are theoretically expected to be equal (Fisher, 1930). If one sex is rare, then, since every offspring has a mother and a father, the average member of that sex will be parent to more offspring than the average member of the other sex. Any gene increasing the frequency of the rarer sex will therefore be selectively advantageous, but decreasingly so as the frequency rises towards 50 per cent. Sex ratios among prosobranchs conform quite well to this prediction, but exceptions may occur among sequential hermaphrodites.

Figure 6.8: Size Frequency Distributions of Male (Open Bars) and Female (Solid Bars) *Littorina saxatilis* among a Total Sample of 259 Snails from a Saltmarsh Population. The distribution of males is centred slightly to the left of that of females. After Hughes and Roberts (1980a)

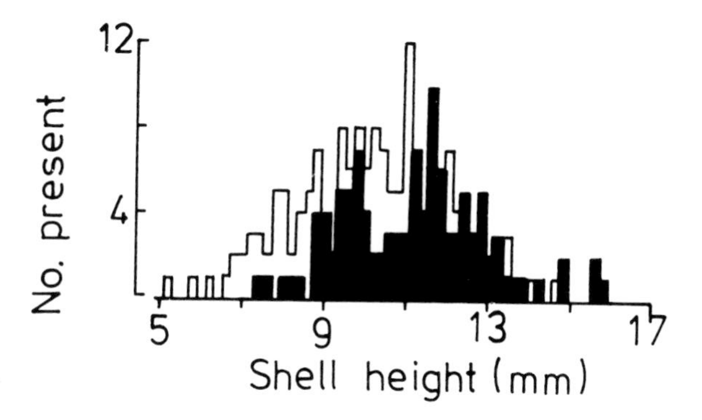

Figure 6.9: Reproductive Effort (P_r/A) Increases as *Littorina littorea* Grows Larger and Older. As estimated from the drop in mean energy content after spawning, the reproductive effort of males is less than that of similarly sized females. After Grahame (1973)

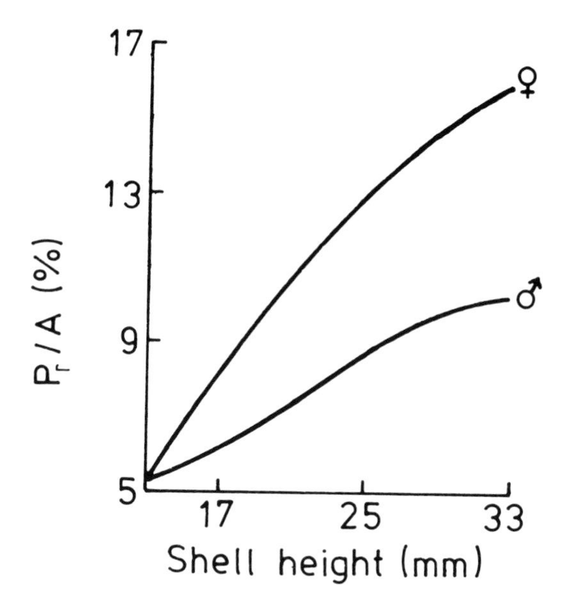

Figure 6.10: Protandric Sequential Hermaphroditism is Common Among Species of *Patella*. Discounting individuals with undifferentiated gonads (N = neuter), the ratio of males to females in the population tends to approach unity. The transition of an individual from male to female is usually total and stages with differentiated gonadal tissue of both sexes are very rare, except in *P. oculus*. After Branch (1981)

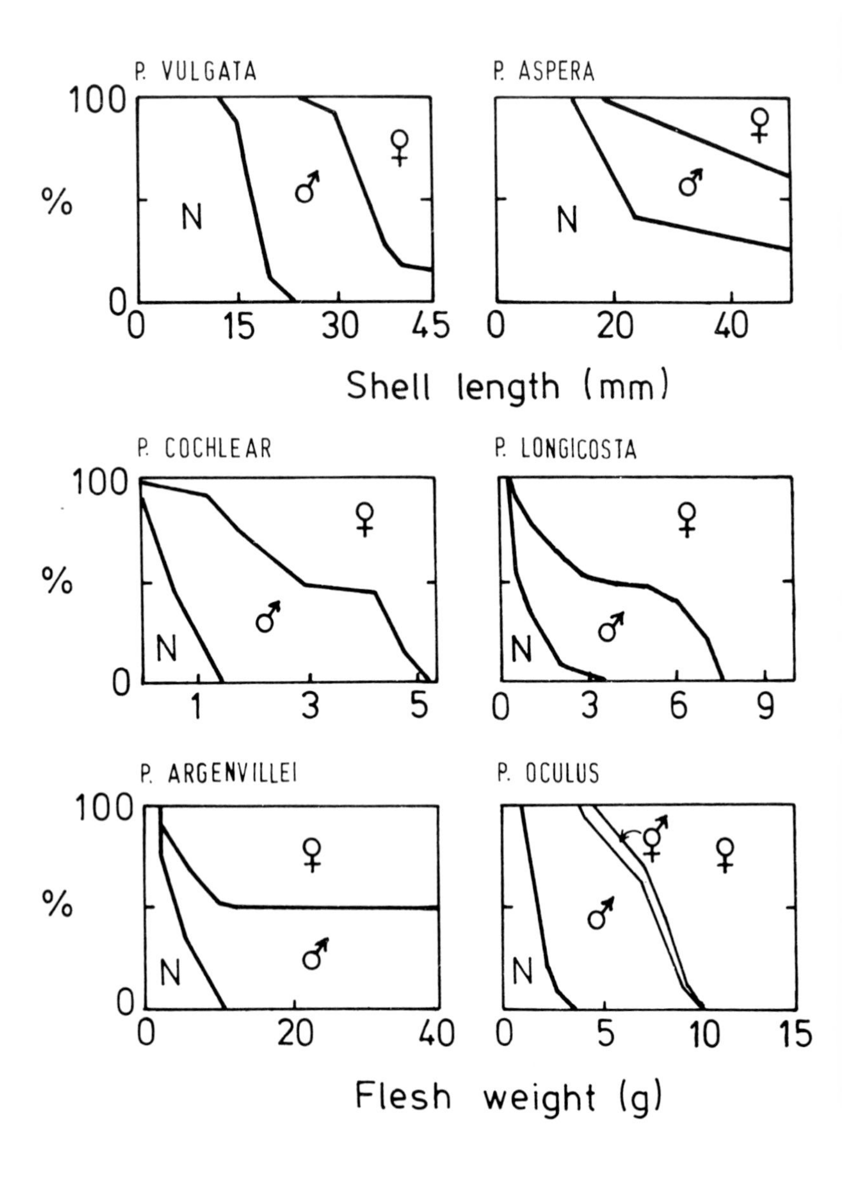

6.4 Hermaphroditism

6.4.1 Simultaneous Hermaphroditism

Simultaneous hermaphroditism is characteristic of opisthobranchs, which exchange sperm at copulation. Because two sets of reproductive organs are developed and maintained by simultaneous hermaphrodites, less reproductive energy is available for gametogenesis and so their potential fecundity must be less than it would be if they were gonochoristic (Heath, 1977). On the other hand, since any pair of mature individuals can copulate, simultaneous hermaphroditism increases the chance of fertilisation and reduces the cost of searching for mates. This is especially advantageous when meetings are rare. Opisthobranchs often occur at low densities compared with many prosobranchs, yet some prosobranchs occur at extremely low densities but are gonochoristic. The universal occurrence of simultaneous hermaphroditism among opisthobranchs may have had an ancestral, low-density advantage, but has become fixed during radiation of the taxon. If so, the energetic penalty of simultaneous hermaphroditism in opisthobranchs cannot be very great.

At extremely low densities, the chance of cross-fertilisation may be vanishingly small and self-fertilisation remains the only possibility. The majority of opisthobranchs for which data are available do not self-fertilise, but this does occur regularly in the ectoparasitic *Odostomia modesta* (Robertson, 1966) and in certain sacoglossans when these are isolated from potential mates (Grahame, 1969).

6.4.2 Sequential Hermaphroditism

In several patellid limpets, males predominate among the smaller mature individuals and females among the larger ones. Size-frequency data and histological material strongly suggest that in these species most individuals mature first as males, changing to females (protandry) when they grow larger (Branch, 1981). Direct tests of this inference by *in vivo* sampling of the gonadal tissue (Wright and Lindberg, 1979) is, however, needed to clarify most cases. Usually, a proportion of the males do not change sex but grow equally as large as the females (Figure 6.10), prompting the question of why sex change is advantageous.

Warner (1975) showed theoretically that sequential hermaphroditism is advantageous when the sexes function best at different sizes. If males function better when small and, or, females function better when large, then protandry is advantageous. Female limpets certainly must exceed a minimum size in order to produce a viable quantity of planktonic larvae (section 6.6), but because fertilisation is external, both sexes gain fecundity as they grow and on this account large size would seem advantageous to males also. On the other hand, sperm is cheaper to produce in quantity than eggs, and small individuals perhaps can function effectively as males without detracting too much energy from somatic growth. If males are sufficiently plentiful not to limit fertilisation, reproductive value will be increased at larger sizes by changing from male to female. This is because all eggs are potential offspring and a greater proportion of female than of male reproductive resources is therefore likely to yield recruits to the next generation. The mixture of gonochoristic males and sequential hermaphrodites presumably represents the counterbalancing advantages of increased male fecundity associated with large size and of becoming female when large.

Protandry is characteristic of most species of *Crepidula*, among which it has clear advantages (Hoagland, 1978). *Crepidula fornicata* is a filter-feeder (section 2.1.4) attaching itself to hard

Figure 6.11: *Crepidula fornicata* Forms Stacks by the Attraction of Settling Larvae to Adults. Older, larger individuals are female, younger ones are male. I — Intermediate stage changing from male to female. After Hoagland (1978)

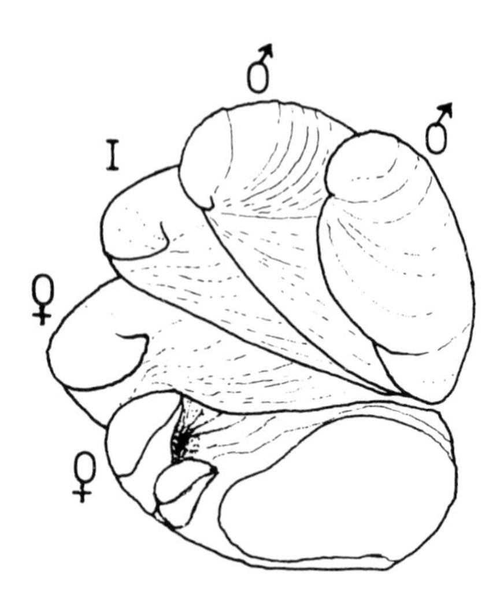

objects, such as oyster shells, in muddy bays. Substratum is seldom limiting, because individuals pile up on one another to form stacks (Figure 6.11). Planktonic larvae are attracted to adults and settle among them, developing first as males but becoming female on attaining a certain size. Thus the older, larger individuals in a stack are female, and the younger, smaller individuals male. The length of the penis increases with distance from the females, allowing all males to reach a female. Moreover, the smaller males are more mobile than the larger ones and can move within reach of most females. Consequently, fecundity in males does not increase with size. Female fecundity on the other hand increases with gonadal volume. Males therefore function best when small and females when large, making protandry advantageous. If larvae fail to encounter adults, they will eventually settle alone, but develop into females that subsequently attract other larvae which will first develop into males and provide the opportunity for fertilisation. There is little risk of inbreeding, because settling planktonic larvae are unlikely to be of the same brood. The size and therefore age at which males become female increases with the density of females (Figure 6.12). Apparently there is an underlying physiological propensity to develop as a female but this is suppressed by a pheromone emanating from established females, making adjacent young develop into males.

Crepidula convexa is sympatric with *C. fornicata* but is generally smaller and more mobile. It does not form stacks and males are paired only temporarily with females. Young are brooded until they hatch at the crawling stage, but there is no attraction of newly hatched juveniles to the adults, thus promoting outbreeding. The advantages of male mobility and female largeness make protandry advantageous, as in *C. fornicata*, but sex change is not influenced by the presence of females. Since the age or size at sex change is genetically fixed, the sex ratio of populations of *C. convexa* is determined by recruitment and mortality and is therefore liable to fluctuate, although direct development will tend to maintain fairly even recruitment. By contrast, recruitment of *C. fornicata* is erratic owing to the vagaries of planktonic development and would cause wide fluctuations in the sex ratio if the age at sex change were genetically fixed. The density-dependent adjustment of sex change, however, maintains a sex ratio close to unity (Figure 6.13).

Protandry, presumably associated with similar advantages to those in *Crepidula*, occurs in parasitic mesogastropods (section

Figure 6.12: As they grow, *Crepidula fornicata* Change from Male to Female,but Tend to Do so at Larger Body Sizes when Females Are More Numerous. The data are means for different populations. After Hoagland (1978)

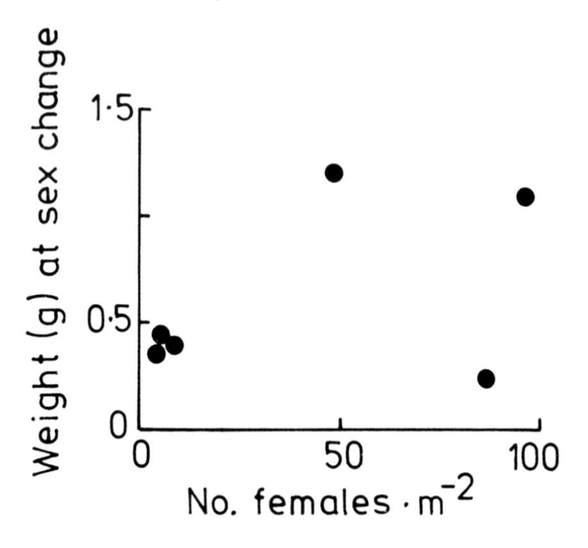

Figure 6.13: The Proportion of Males in a Population of *Crepidula fornicata* is Approximately 50% at Moderate Densities, but there is a Tendency of the Sex Ratio to Be Biased towards Males at High Population Densities. After Hoagland (1978)

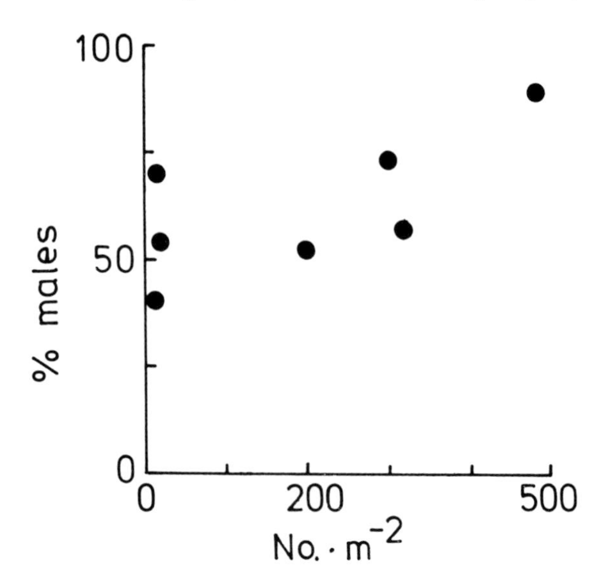

2.1.8). In *Stilifer linckiae*, the small male does not change sex if a female is present. As starfish hosts rarely contain more than a few *S. linckiae*, the suppression of sex change ensures a co-occurrence of both sexes (Lützen, 1972). Cases of protandry have also been recorded in the Ianthinidae, the Scalidae, and in the mesogastropod *Lora kurricula* (reviewed in Hoagland, 1978).

6.5 Parental Investment

6.5.1 Egg Capsules and Jelly Masses

Parents invest energy and time and may take risks on behalf of their young: the allocation per offspring varying greatly among species and even among the progeny within a clutch in certain cases. Many archaeogastropods, notably patellid limpets, spawn myriads of tiny eggs, which, after external fertilisation, develop into veliger larvae that feed on phytoplankton for several weeks (planktotrophic development) before settlement and metamorphosis. Externally fertilising archaeogastropods therefore expend minimal parental investment.

Most other gastropods enclose the eggs in jelly masses or in capsules secreted by albumin and capsule glands in the terminal region of the reproductive tract. *Littorina neritoides* encloses each egg in a delicate capsule that remains buoyant in the plankton until the veligers hatch and lead a free-swimming existence. A wide variety of meso- and neogastropods enclose batches of eggs in substantial proteinaceous capsules cemented to the substratum (Plate 6.1).

Dogwhelks aggregate during the winter and communally deposit their egg capsules in large clumps. *Nucella lamellosa* on the Pacific coast of North America, for example, forms breeding groups of up to 200 females depositing several thousand capsules in one mass (Spight, 1975). Close packing of the capsules traps interstitial water and reduces the total surface area to volume ratio, reducing desiccation and the effect of wave impact. Aggregation precludes foraging, so the dogwhelks starve throughout the spawning season and lose weight. About 19 per cent of this weight is accounted for by catabolism, 45 per cent by the production of eggs and intracapsular fluid, and 36 per cent by production of the capsules themselves (Stickle, 1975). Capsules therefore represent

Plate 6.1: *Buccinum undatum* Secretes Proteinaceous Capsules, Each Containing a Clutch of Eggs. The capsules are attached to one another in a dome-shaped mass that is cemented to the substratum subtidally. Interconnecting spaces around the capsules facilitate ventilation of those beneath the surface of the mass. Deposition of the egg capsules in a cluster may enable the whelk to use small items of hard substratum, such as empty shells, lying on soft sediments or it may confer some protection against predation. Each capsule is about 10 mm long. Several whelks may deposit capsules in the same mass. Photograph by E.W. Pritchard

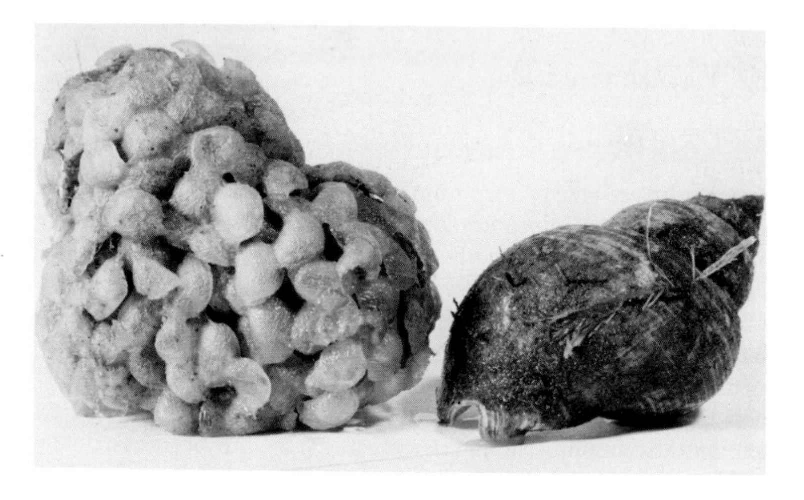

about a third of parental investment in *N. lamellosa* and so must have a considerable selective advantage.

Capsules are partially resistant to environmental fluctuations. The capsular wall of *Nucella lapillus* is readily permeable to water and sodium chloride but much less so to the small organic molecules in the intracapsular fluid: thus, when at equilibrium with water of 150 mosm litre^{-1}, the osmotic concentration of the intracapsular fluid remains about 25-30 mosm litre^{-1} above ambient (Pechenik, 1983).

Nevertheless, eggs of *N. lapillus* and *N. lamellosa* frequently die within their capsules (Spight, 1976), which are not, therefore, very resistant to environmental stress. Nor are capsules or jelly masses particularly resistant to infection by micro-organisms: capsules secreted by *Urosalpinx cinerea* do not always protect the embryos against fungal attack (Ganaros, 1957), and the jelly masses of cerithiids (Houbrick, 1973) and nudibranchs (Hadfield, 1963) sometimes fail to prevent bacterial and protozoan infestation.

Capsules and jelly may offer protection against predation. The egg capsules of *Ilyanassa obsoleta* spend over a week attached to the substratum during which time over 50 per cent may be eaten accidentally by grazing *Littorina littorea*, but the capsules resist the digestive enzymes, enabling the embryos to survive passage through the gut (Brenchley, 1982). Carnivores, however, are more likely than grazers to damage egg capsules; significant proportions of dogwhelk egg capsules are opened by predators (Spight, 1975) and in general capsules appear not to be very effective against predation (Pechenik, 1979).

Intracapsular fluid may nourish embryos: intracapsular protein is pinocytosed by kidney epithelial cells of embryonic *Serlesia dira* (Rivest, cited by Kohn, 1983). But embryos of *N. lapillus* are provisioned with only about 1.1 μl of fluid and they lose weight during development (Pechenik *et al.*, 1984); and in *Conus pennaceus* intracellular fluid represents only 10 per cent of the energy content of the capsule (Perron, 1981), suggesting that nutrition is not generally an important function of encapsulation. What then is the main advantage of this form of parental investment?

Non-motile eggs and embryos, although small, would tend to sink to the sea-bed where risks of predation are probably much greater even than they are in the plankton. Sediments abound in predatory foraminiferans, turbellarians, copepods and deposit-feeding polychaetes, molluscs and echinoderms, and rocks are often carpeted with filter-feeding animals. Photopositive swimming behaviour, characteristic of newly hatched planktonic larvae, may have evolved not only because it lifts them into the water column for dispersal or feeding, but also because it removes them from the vicinity of benthic animals that might engulf them (Crisp, 1974). A principal advantage of encapsulation, therefore, may be to protect eggs and embryos from benthic predators (Pechenik, 1979).

Encapsulation also facilitates the ventilation of benthic embryos, necessary for adequate gaseous exchange. Some animals, such as herring, lay sticky eggs that adhere to one another and to the substratum in a dense mass. Gaseous exchange by diffusion is restricted for embryos deeply embedded in the mass and this may severely retard their developmental rate or even kill them. However, when embryos are embedded in a matrix of jelly, as with a variety of prosobranchs and most opisthobranchs, they can be spaced out from one another, so increasing their ventilation (Strathmann and Chaffee, 1984).

But even when embedded in jelly, central embryos can suffer reduced ventilation if the mass is globular. The bullomorph *Melanochlamys diomedea* lays thick gelatinous egg masses in which the central embryos are less well ventilated, and so develop more slowly than the peripheral ones. This causes asynchronous development among the embryos, but synchrony can be induced experimentally by cutting the egg mass into small pieces (Chaffee and Strathmann, 1984). Another bullomorph, *Haminoea vesicula*, from the same habitat lays a thin gelatinous egg ribbon which avoids problems of ventilation and allows synchronous development of the embryos.

The egg capsules produced by many meso- and neogastropods may promote ventilation even further than do jelly masses because the intracapsular fluid can be circulated. If the fluid is well mixed, the number of embryos could increase in proportion to the surface area of the capsule, whereas because of limitations on diffusion, they could increase only in proportion to the radius of spherical jelly masses (Strathmann and Chaffee, 1984). The number of embryos per capsule is correlated with capsular surface area in *Conus pennaceus* (Perron and Corpuz, 1982).

6.5.2 Brooding

Some gastropods brood their offspring, protecting them from environmental hazards and predation. Brooding generally occurs more frequently among smaller than among larger species, although the reverse may be true of certain trochaceans (Strathmann and Strathmann, 1982). Theoretical explanations for this trend have invoked two possibilities: first, the incapacity of large adults to accommodate the large numbers of young resulting from the allometric relationships between fecundity and body size; secondly, the ability of large adults to produce young in sufficient quantities to offset the high mortality rates suffered by independent larvae while gaining advantages of dispersal and, since large adults tend to be long-lived, of achieving high reproductive success in years conducive to larval survival and settlement. These putative relationships are, as yet, little understood, and the tendency for brooding to be associated with small adult size may result from several different selective forces in different taxa (Strathmann and Strathmann, 1982).

Clear advantages of brooding, irrespective of adult body size, may be seen among littorinids. *Littorina arcana* secretes jelly around each clutch of eggs, cementing them to rocks; but the jelly gland of the sibling species *L. saxatilis* is modified into a brood chamber in which the eggs are retained until hatched (Figure 6.14A). Whether embryos derive nutrition from the brood chamber is not known, but retention of the eggs reduces the number that can be laid in a season compared with the oviparous species, and is a cost of this type of parental investment. The reduced fecundity is compensated by the ability to rear young in places too harsh for external egg masses, such as occur high on rocky shores and on silty saltmarshes (Hughes and Roberts, 1981).

Pteraeolidia ianthina wraps its body round the egg mass while clinging to the substratum (Rose and Hoegh-Guldberg, 1982), and *Bullia melanostoma* holds the eggs in its foot (Figure 6.14B). Parental investment of this nature may be costly because it prevents foraging or renders the parent more vulnerable to predators. *Neptunea pribiloffensis*, inhabiting Alaskan rocky shores, deposits large egg masses that are readily consumed by sea urchins if left unguarded. The egg masses, however, are usually laid next to the large sea anemone, *Tealia crassicornis*, which consumes any urchins coming within range. Not only does *N. pribiloffensis* use *T. crassicornis* as a 'baby-sitter' to protect its eggs, but also the anemone may benefit from the association if urchins are attracted to the 'bait' provided by the eggs (Shimek, 1981).

6.5.3 Provision of Yolk and the Evolution of Nurse-eggs

A major form of parental investment among gastropods is the endowment of embryos with yolk. Larger eggs containing more yolk hatch into large juveniles, which generally survive better than smaller ones (Spight, 1976). However, fewer larger eggs can be produced per unit of reproductive effort and they develop more slowly (Figure 6.15), partly because yolk retards cleavage.

Gastropods exploit two methods of reducing the risk of mortality associated with prolonged development. First, capsules enclosing larger eggs may be made stronger (Figure 6.16). *Conus vexillum* produces eggs of 140 μm diameter, allocating 20 per cent of spawned energy to capsular material, whereas *C. pennaceus* produces eggs of 500 μm diameter, allocating 47 per cent to capsular material, equivalent to a decrease in fecundity of 34 per cent compared with *C. vexillum* (Perron, 1981). Given a fixed amount of available energy (Sibly and Calow, 1982), optimal capsular thickness is likely to be a compromise between added

Figure 6.14: A. *Littorina saxatilis* (14 mm) Broods its Embryos to the Crawling Stage. The shell has been cut away to reveal the brood-chamber, which is a modified section of the reproductive tract. B. *Bullia melanostoma* (40 mm) Inhabits Surf-Beaches of India. It protects its eggs by enveloping them in the folds of the foot. After Ansell and Trevallion (1970)

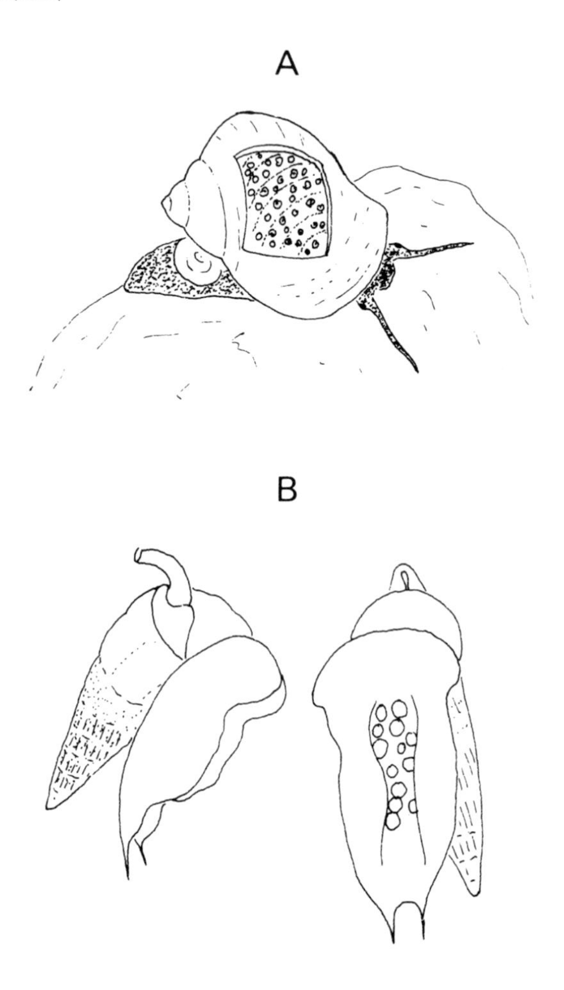

protection, optimal size and number of embryos per capsule. Secondly, prolonged development may be avoided by producing small eggs and provisioning them with extra-embryonic yolk (Spight, 1975).

Deaths are frequent among developing embryos. Mortalities of 20-50 per cent have been recorded among encapsulated muricid

Figure 6.15: A. Development Time to Hatching Is Positively Correlated with Egg Size among Species of *Conus*. After Perron (1981). B. Development Time to Hatching is Positively Correlated with Egg Size among Nudibranchs, with the Result that on Average those with Direct Development Take the Longest Time to Hatch and those with Planktotrophic Development the Least. After Todd and Doyle (1981)

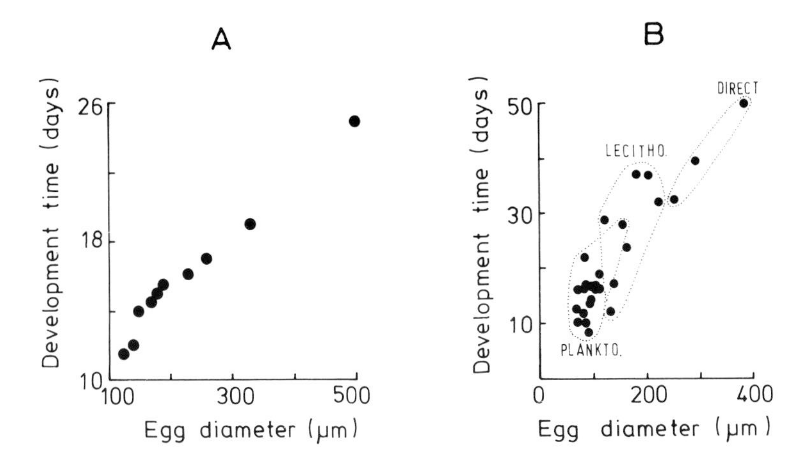

Figure 6.16: Among Species of *Conus*, Capsular Strength is Positively Correlated with Development Time to Hatching (and Hence with Egg Size). Capsular investment is expressed as (energy content)/(energy content of capsule + energy content of eggs) × 100. After Perron (1981)

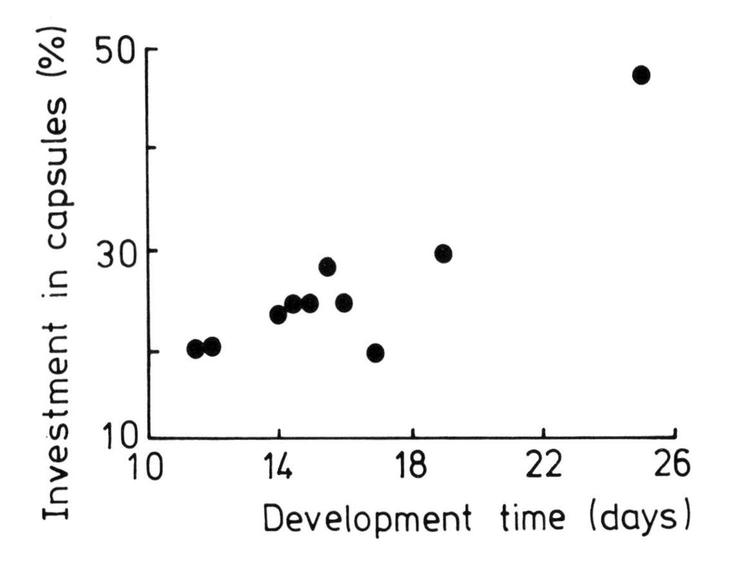

embryos (Spight, 1975) and in most cases the abortive individuals are eventually eaten by their siblings. During about the last third of intracapsular development, embryos possess velar lobes with which they are able to enfold and fragment defunct siblings (Fioroni, 1966). Consumption of moribund embryos not only helps to keep the capsule free of microbial infection, but also supplies the surviving embryos with extra yolk. Embryos that have consumed yolk develop quickly, commensurate with their small size, but owing to the nutritional subsidy they hatch at a more advanced stage of development.

As a rule, small embryos without substantial yolk reserves hatch as planktotrophic veligers that depend on planktonic food to supply the energy needed for continued development and survival. Those with intermediate yolk reserves hatch as lecithotrophic veligers that do not require planktonic food but metamorphose within a few days or even a few hours after hatching. Those with large yolk reserves complete metamorphosis within the capsule (direct development) and emerge as crawling young.

Usually each species has a characteristic form of development corresponding to the endowment of yolk to the embryos, but in certain species the endowment seems to involve an element of chance. In certain vermetids (Hadfield *et al.*, 1972) and in *Murex incarnatus* (Gohar and Eisawy, 1967), embryos in some capsules develop directly into crawling young, whereas in others of the same brood they emerge as planktonic veligers, depending on whether or not they have eaten abortive embryos. In other cases, the provision of extra-embryonic yolk is more controlled, whereby a fraction of the eggs always abort at some stage in development, acting as 'nurse-eggs' for the survivors.

Nurse-eggs are prevalent among gastropods consistently undergoing direct development, but in some species the proportion of nurse-eggs per capsule varies according to environmental conditions, leading to direct development when the proportion is high and to planktonic veligers when it is low. For example, at certain localities most eggs of *Natica catena* are viable and hatch as veligers, but elsewhere many of them are infertile and are consumed by the viable embryos, which hatch as crawling young (Thorson, 1950). *Elysia cauze* provisions embryos with a yolk ribbon, analogous to nurse-eggs, that ramifies through the egg mass. Development varies from planktotrophy in the spring, lecithotrophy in the summer, to direct development in the autumn, as the amount of yolk ribbon is altered (Clark *et al.*, 1979).

Differential parental investment among populations may be genetically determined in some species. *Elysia chlorotica* on New England shores has planktotrophic development in certain populations and direct development in others. When individuals from the two types of population are hybridised, the F-$_1$ egg masses contain either small eggs with planktotrophic development or larger ones with direct development, whereas the F-$_2$ egg masses have intermediate-sized eggs with planktotrophic development (West *et al.*, 1984).

6.6 Larval Developmental Mode

The relative advantages of planktonic and direct development have long been debated (Thorson, 1946; Vance, 1973), but difficulties of measuring the survivorship and dispersal of larvae have stifled any practical test of the ideas, so logic and circumstantial evidence remain as the only means of appraisal. Dispersal of planktonic larvae could be selectively advantageous when adults have limited powers of dispersal and exploit scattered patches of habitat that vary temporally but randomly in quality (Crisp, 1976). Some of the dispersing larvae will probably colonise newly favourable patches, increasing parental contribution to future generations, while insuring against deterioration of the original site. Larvae, however, drift passively without any means of controlling their direction, and many of them by chance will never encounter a suitable place for settlement. On the other hand, when ready to metamorphose, larvae become photonegative, bringing them close to the substratum, which they can explore by touch and olfaction. If the substratum is suitable, the larvae will settle, but if not they will delay settlement for a time, during which a suitable place may be encountered.

This 'delay period' is limited by energy reserves in lecithotrophic larvae but not in planktotrophic ones, among which it shows great interspecific variation. Pechenik (1980) hypothesised that larval development has a genetically fixed end point and that the potential length of the delay period is governed by the rate at which larvae progress towards it. For example, larval *Ilyanassa obsoleta* have lower absorption efficiencies and higher respiration rates than larval *Crepidula fornicata*, causing them to progress more slowly through their developmental programme and so imparting greater delaying capabilities.

Remarkable delaying capabilities are found in certain gastropod larvae. Veligers of *Aplysia juliana* stop growing some 30 days after hatching from the egg mass, having reached the 'competent' stage at which they will respond to the presence of *Ulva*, the preferred food of the adult, by settlement and metamorphosis. In the absence of *Ulva*, the larvae maintain a constant tissue and shell mass; swimming, feeding and remaining competent to metamorphose for over 200 days (Kempf, 1981). An exponentially decreasing survivorship curve (Figure 6.17) throughout the delay period suggests that extrinsic mortality factors account for most larval deaths, but eventually brown pigment accumulates in the larval foot and mantle, indicating senescence, perhaps equivalent to Pechenik's 'developmental end point'. Not surprisingly, *A. juliana* is circumtropical in distribution: currents will carry a drifting larva from Japan to the Hawaiian archipelago in about 77 days, well within the delay period of *A. juliana*.

Veligers able to remain planktonic for a long time (teleplanic) probably traverse the oceans quite commonly. For example, the Caribbean species *Cymatium parthenopeum* produces teleplanic veligers in such quantity that 1.3×10^{12} of them are estimated to

Figure 6.17: The Survivorship Curve of Veliger Larvae of *Aplysia juliana* Hatched in the Laboratory Is Approximately Exponential, Implying a Constant Instantaneous Rate of Mortality. In this culture, three veligers were still alive 316 days after hatching. After Kempf (1981)

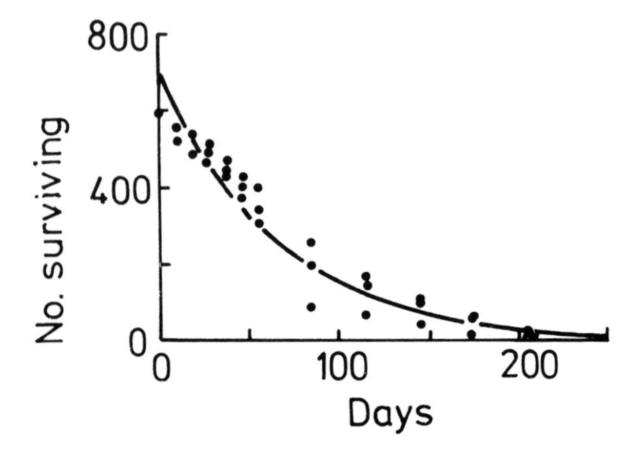

cross the North Atlantic each year, having successfully colonised the Azores and occasionally metamorphosed off Ireland (Scheltema, 1978). Because species with teleplanic larvae have wide geographical ranges, they are probably generalists with wide environmental tolerances. If so, they will be less susceptible to extinction and should have a longer fossil record than species with poorer dispersal capabilities. Fossil shells with protoconchs similar in shape to those of related extant teleplanic species would probably themselves have been teleplanic. Although exceptions occur, species with non-planktonic larvae tend to have protoconchs with fewer, wider whorls than those with planktotrophic larvae (Figure 6.18) and of the latter, teleplanic forms tend to have a convoluted rim of the protoconch (Figure 6.18), which accommodates the extended velar lobes. Based on such comparisons, Scheltema (1978) inferred that members of the Bursidae and Cymatiidae with long fossil records were teleplanic.

Figure 6.18: According to Thorson's 'apex theory', planktotrophic veliger larvae tend to have narrow, many-whorled protoconchs, whereas lecithotrophic or direct-developing veligers of related species have more rounded, wide paucispiral protoconchs. This is illustrated by the protoconchs of *Rissoa guerini* (upper left), which is planktotrophic, and of the closely related *Barleeia rubra* (upper right), which is non-planktotrophic. Both protoconchs are about 1 mm in diameter. After Jablonski and Lutz (1980). Teleplanic veligers, which live for long periods in the plankton, often have a folded (sinusigera) lip on the aperture that accommodates the extended velar lobes of the larva. This is illustrated by the protoconch of *Thais haemastoma* (lower centre, 1.4 mm height). After Scheltema (1978)

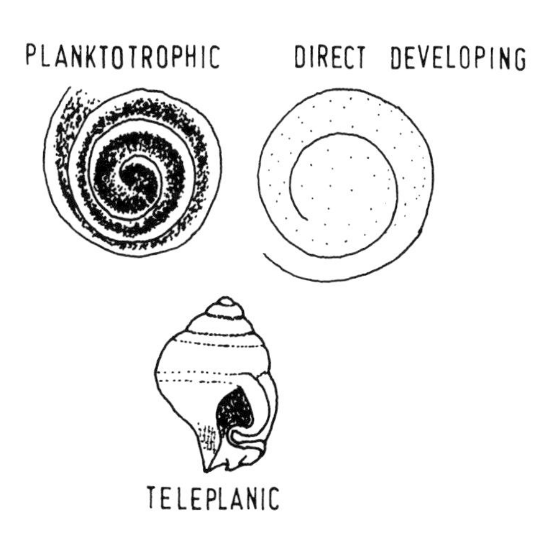

The maximum delay period of *Aplysia juliana* is at least 6.7 times longer than the developmental time required to reach metamorphosing competence, suggesting that prolonged planktonic life serves only to increase the chance of eventually reaching a suitable place for settlement. This is not to say that long delay periods are advantageous by increasing the mean distance dispersed. Although dispersal is important (Crisp, 1976), diminishing returns are likely to accrue to larvae spreading beyond certain limits, which will be greater when patches of habitat are more variable (Figure 6.19). Beyond these limits, it is the length of the delay, not the distance from the origin, that increases the chance of successful settlement.

Among dispersing larvae, planktotrophy may have two advantages over lecithotrophy. First, because it feeds to sustain itself, the planktotrophic larva can delay settlement for much longer periods than the lecithotrophic larva, which starves after exhausting its nutrient reserves. Secondly, because planktotrophic larvae are not

Figure 6.19: According to One Mathematical Model, the Advantage of Larval Dispersal Rises Asymptotically (Law of Diminishing Returns) with Increasing Amounts of Spread from the Origin. The advantage is potentially much greater when the habitat is spatially or temporally highly variable. From Palmer and Strathmann (1981)

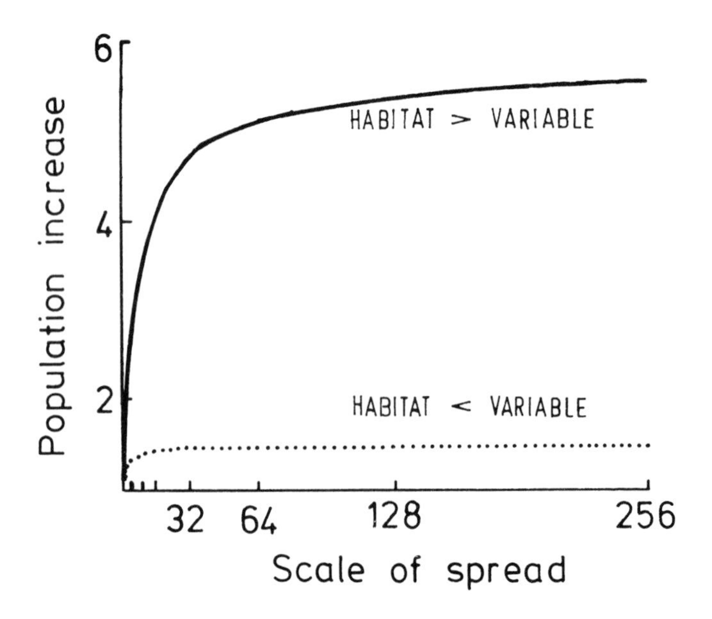

provisioned with nutrients, they are smaller and can be produced in larger quantities. But there are two disadvantages of planktotrophy. First, the small planktotrophic larva hatches at an earlier stage in development and requires a longer period to reach competence, during which time it cannot settle but is exposed to mortality and may be carried away from suitable habitat. Secondly, the planktotrophic larva may find a shortage of food. This risk, however, is reduced when spawning coincides with plankton 'blooms' (Himmelman, 1975).

Todd and Doyle (1981) have suggested that, at least in *Onchidoris bilamellata*, planktotrophic development is advantageous not primarily because it maximises fecundity or increases dispersability, but because it spans the gap between the optimum times for spawning and settlement ('settlement-timing' hypothesis). On local shores, growth and survivorship of *O. bilamellata* interact to give a maximum reproductive value (section 6.2.1) in late February and this is when spawning takes place. Newly metamorphosed *O. bilamellata* are about 0.5 mm long and feed only on barnacle spat, being incapable of handling the larger barnacles eaten by adults. Settlement of *O. bilamellata* larvae therefore must coincide with that of *Semibalanus balanoides* in April–May. Planktotrophic development can be prolonged over the 13-week period between optimum times for spawning and settlement, whereas lecithotrophic and direct development would proceed too quickly, causing settlement to fall short of the critical time (Figure 6.20). Because spawning at a certain time of year is associated with a particular range of temperature, altering the egg size is a simple way of adjusting the period from spawning to settlement. This will only work, however, over a seasonal trend of increasing temperatures such as occur in the spring of temperate regions. Larval developmental rate has a Q_{10} of about 2 and later-produced eggs experiencing higher temperatures will develop quicker, telescoping the progress of staggered clutches so that settlement is synchronised. If spawning occurred when temperatures were falling in the autumn, later clutches would take longer to develop and demographic synchrony would be destroyed, with adverse consequences to mating and spawning.

Physiological constraints, however, may leave no option on developmental mode. Larvae must hatch beyond a minimum size in order to be viable and very small animals may have insufficient resources to produce the large number of eggs necessary to

Figure 6.20: Settlement-Timing Hypothesis for the Optimal Mode of Larval Development. *Onchidoris bilamellata* spawns in mid-January, when its reproductive value is at a maximum. The long planktonic phase of the planktotrophic larvae delays settlement until early May, when barnacle spat have settled and provide food for the young *O. bilamellata*. If *O. bilamellata* were to have larvae with lecithotrophic or direct development, then either settlement would occur too early (as indicated on the figure) and miss the spatfall of barnacles, or spawning would have to be delayed beyond the time of maximum reproductive value. Solid bar = pre-hatching phase, open bar = planktonic phase. After Todd and Doyle (1981)

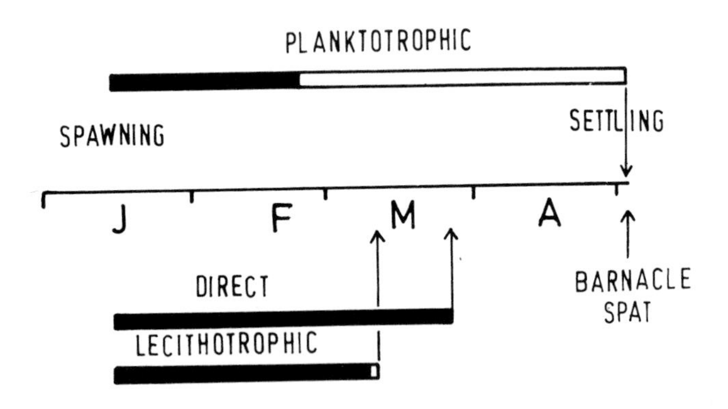

counterbalance the enormous losses in the plankton. Accordingly, tiny gastropods such as *Lacuna parva* (Ockelmann and Nielsen, 1981) and *Littorina neglecta* (Hannaford-Ellis, 1983) devote their limited resources to the production of a few well-endowed directly developing eggs, which have a relatively high chance of survival.

6.7 Hormonal Control of Reproduction

The environment is rarely constant for any length of time and some phases of its variation will favour the survival and growth of progeny more than others. Reproductive cycles that allow progeny to be released during these favourable periods will therefore be selectively advantageous. Before eggs or young are released into the environment, a complex, protracted series of events takes place involving the mobilisation of stored nutrients, gametogenesis, gamete maturation, fertilisation and perhaps encapsulation or brooding. Each of these events must be appropriately timed if the gastropod is to reach spawning condition when optimum environmental conditions prevail.

Probably all phases of the reproductive cycle are under hormonal control (Le Gall and Streiff, 1975), and increasing evidence suggests that reproductive hormone levels in molluscs are influenced by environmental factors such as photoperiod, temperature and nutritional state. A well-documented example comes from work on chitons, close relatives of the gastropods. A decrease in temperature is required to initiate gametogenesis in the chiton *Katherina tunicata* and an increase in temperature is needed to complete it, whereas spawning is stimulated by high phytoplankton levels coinciding with ambient temperatures above a threshold of 12°C (Himmelman, 1978). Similar phenomena may be expected among gastropods. Gametogenesis proceeds at temperatures below those necessary for spawning in *Nassarius* spp. (Pechenik, 1978a,b), in several Australian prosobranchs (Underwood, 1974) and in some opisthobranchs (Thompson, 1958).

It is not yet known to what extent the endocrine system is stimulated directly by environmental factors or indirectly via the nervous system, but there is clear evidence of some interaction between the two systems. The nervous and the endocrine control of molluscan reproduction have been studied most intensively in *Octopus*, in the freshwater snail *Lymnaea stagnalis* and in *Aplysia* (Joosse, 1979; Lever and Boer, 1983). In these normally semelparous animals, growth and reproduction appear to be antagonistic processes. Reproductive development of adult *Octopus* is controlled by endocrine activity of the optic gland, itself regulated by the nervous system in response to environmental factors. Activity of the optic gland in young, growing individuals is suppressed by neuronal control from the frontal lobe. Severing the nerves from the frontal lobe allows precocious activation of the optic gland, with consequent sexual development and reduced growth. An analogous neuroendocrinal mechanism controlling growth and reproduction in *Lymnaea stagnalis* involves endocrinal activity of 'light green cells' (so called because of their staining properties in histological preparations) situated within the cerebral ganglia, together with activity of 'lateral lobes' on the cerebral ganglia. The 'light green cells' secrete a growth-promoting hormone, while activity of the lateral lobes modulates this neurosecretion, with the effect that more resources are available for reproduction.

The generality of these reproductive responses to environmental conditions, mediated by the neuroendocrine system, remains to be explored, but differences must be expected in relation to the various patterns of life history (Runham, in press). For example, there are apparently large differences in the effect of starvation on reproduction. Starved iteroparous chitons continue to spawn, but at a reduced rate, utilising resources stored before the onset of reproductive activity. In contrast, *Lymnaea stagnalis* is normally a semelparous species which spawns continually after reaching sexual maturity, converting all non-respired assimilated energy into eggs (Dogterom *et al.*, 1985). It does not, however, store any nutrients and therefore is incapable of sustaining egg production when starved. There is also considerable variation, among iteroparous species, in the fate of the gonad between successive seasons. On Australian shores *Nodilittorina pyramidalis* completely resorbs unshed oocytes after spawning and the gonad becomes quiescent, whereas *Subninella undulata* stores unshed oocytes until the next spawning season and the gonad continues vitellogenesis (Underwood, 1974).

In addition to controlling the reproductive cycle, hormones also control the expression of gender. As sex chromosomes are apparently absent in molluscs, each individual has the potential to develop into a male or a female. Organ culture experiments have shown that abnormal hormone levels induce penis formation in female *Ocenebra erinacea* (Féral, 1979), and the development of a vestigial penis by some females is a natural phenomenon in *Urosalpinx cinerea* (Hall and Feng, 1976) and *Ilyanassa obsoleta* (Smith, 1981).

Such a labile mechanism for determining gender is conducive to the evolution of hermaphroditism, which evidently has occurred independently in numerous molluscan lineages (section 6.4). In view of this independent evolution, it is not surprising that the mechanism of sex determination varies among taxa. A male-differentiating hormone is produced by neurosecretory activity of the cerebral ganglia in *Calyptraea* spp. but not in *Patella vulgata.* The latter species, on the other hand, produces a hormone in its tentacles which inhibits spermatogenesis but which has not been found in *Calyptraea* spp. (Le Gall and Streiff, 1975). Full details of the hormonal basis of sex determination in molluscs have yet to be elucidated.

The neurohormonal control of growth, gender and reproduction offers a potentially useful system for experimentally testing theoretical predictions about life histories, sex ratio, sexual investments and reproductive effort (see other sections of this chapter). If the variables considered by these theories, such as the age at first sexual maturity, age of sex reversal, or the number of reproductive episodes, could be controlled under standard laboratory conditions, then much more direct and therefore more convincing tests of the predictions could be made than has been possible hitherto. At the moment there are considerable technical difficulties in culturing many species in large enough quantities, in rearing large cohorts of developmentally homogeneous subjects and in performing operations on the neuroendocrine organs (Runham, in press). The benefits to be gained from such an approach are so great, however, that progress may be expected in the future.

Indeed considerable progress has recently been made in understanding the biosynthesis, transport and release of molluscan neurohormones (Lever and Boer, 1983). *Aplysia californica* has proved to be particularly useful for this purpose because on the visceral ganglion are two peripheral clusters of large, protruding nerve cells (bag cells) that are surgically accessible and secrete a reproductive hormone (egg-laying hormone, ELH) whose effects are easily recognised. Each cluster contains about 400 bag cells and these secrete at least three peptides including ELH. This peptide has 36 amino acids of known sequence and a molecular weight of 4400. It acts like a neurotransmitter on nerve cells and like a hormone on other organ systems. During the reproductive period about half the protein secreted by the bag cells is ELH. The neurological influences of ELH include excitation and inhibition of specific neurons in the visceral and buccal ganglia. This results in the entrainment of a stereotyped pattern of reproductive behaviour, initiated by the inhibition of locomotion and feeding, followed by head-waving and spawning. The hormonal influences of ELH include stimulation of the smooth muscles of follicles in the ovotestis to initiate egg laying.

In addition to its use as a model for neuroendocrinological studies, the bag-cell complex of *Aplysia* has also provided a good experimental system for examining, by techniques of genetic engineering, the structure, expression and modulation of genes that code for a peptide (ELH) of known behavioural function (Scheller and McAllister, 1983). To do this, a 'library' of the haploid genome was first constructed from the DNA of sperm from a single *Aplysia*. The DNA was digested with an enzyme to break it into short fragments. These fragments were made to attach to a bacteriophage which was then allowed to infect a culture of the bacterium *Escherichia coli*, yielding over a million clones that together would contain and replicate virtually all the fragment types. Messenger RNA, extracted from the bag cells, the visceral ganglion and the digestive gland, was used with the enzyme 'reverse transcriptase' to synthesise DNA identical to the chromosomal DNA from the three sources. Attempts were made to hybridise this radioactively labelled synthetic 'clonal DNA' with the DNA fragments carried by the different clones of *E. coli*. Of 600 000 clones screened in this way, two hybridised with bag-cell cDNA and showed no hybridisation with cDNA from the visceral ganglion or digestive gland. The DNA of one of these two clones was mapped using 'restriction endonucleases' to identify the smallest fragment containing homologous material to the bag-cell cDNA. The corresponding fragment of DNA was found by partial-sequence analysis to be comprised of 108 contiguous nucleotides, which encode the 36 amino acids of ELH. This DNA fragment is now being used to analyse the organisation and transcription of ELH genes and for characterising the protein products.

6.8 Life Histories

Patterns of development, growth and reproduction constitute life history, and the various conceptual treatments of reproductive effort, semelparity versus iteroparity, sex ratio, hermaphroditism, parental investment and larval development considered in the previous sections are all within the domain of life-history theory. Evolutionary ecologists have tried to produce a general theory of life history that will predict the natural permutations and combinations of these phenomena.

Perhaps the most widely used idea is that of '*r*-' and '*K*-selection' originally proposed by MacArthur and Wilson (1967). The essence of MacArthur and Wilson's reasoning is as follows. If organisms live in situations where their populations are repeatedly reduced to low densities, for example by fluctuating environmental conditions, then genotypes conferring higher rates of population increase during intermittent favourable periods will be selectively advantageous. Rapid population expansion enables organisms to exploit temporarily abundant resources or temporarily benign environmental conditions at the expense of less prolific ones. High potential rate of population increase is associated with high reproductive rate and short generation time, both associated with a particular suite of biological characteristics. High reproductive rate requires high fecundity, and short generation time requires rapid development and early sexual maturity. Since resources available for reproduction are limited (section 6.2), high fecundity is associated with high reproductive effort and small progeny. Since weight-specific metabolic rate is inversely proportional to body mass (section 4.2.1), rapid development is associated with small body size. High fecundity coupled with small adult size will depress survivorship and there will therefore be a tendency towards semelparity (section 6.2.4). The opportunistic use of resources will be enhanced by an ability to exploit a wide variety of resources, leading to generalist 'niches'.

Because these properties are advantageous under conditions of unlimited population expansion, MacArthur and Wilson denoted them as characteristic products of '*r*-selection' (Table 6.1), referring to the symbol for unlimited instantaneous per capita rate of population increase in the 'logistic' growth equation. '*r*-selection' may be self-reinforcing, since small size renders organisms more vulnerable to environmental deterioration, making their populations less stable and increasing the advantage of high potential rate of increase (Horn, 1978).

If organisms live in situations where their populations are able to remain about an equilibrium density, close to the 'carrying

Table 6.1: Attributes Theoretically Predicted to Arise from Natural Selection when Populations Are Repeatedly Reduced to Low Densities, with the Result that Rapid Population Increase is Advantageous (*r*-selection) and When they Remain Close to an Equilibrium Density with the Result that Competitive Ability is Advantageous (*K*-selection)

r-selection	*K*-selection
Small body size	Large body size
Early sexual maturity	Delayed sexual maturity
Many, small offspring	Few, well-developed offspring
Brief development	Prolonged development
High reproductive rate	Low reproductive rate
Semelparity	Iteroparity
Short generation time	Long generation time
Opportunistic use of resources (fluctuating population density)	Specialised use of resources (steady population density)

capacity' of the habitat, then genotypes conferring greater competitive ability will be selectively advantageous. Competitive prowess enables organisms to gain more resource, hence scope for production, at the expense of poorer competitors. High competitive ability tends to be associated with biological characteristics opposite to those associated with high potential rate of increase. High parental investment, either in the form of well-provisioned embryos or brooded young, ensures that progeny are relatively large, developmentally advanced and hence stronger competitors when they become dependent on external resources. Because resources for reproduction are limited, high parental investment per offspring will depress fecundity. Large adult size may also increase competitive ability, but it will take the organism longer to grow to the size at sexual maturity. Organisms with a long growth period will benefit from the development of protective mechanisms that increase survivorship, but such investments will concomitantly reduce resources available for growth and reproduction. Moreover, as size increases, weight-specific metabolic rate decreases and the combined result of delayed sexual maturity, reduced scope for production and reduced metabolic turnover rate will be to lengthen generation time. The chances of recruitment in a highly competitive population are likely to be small and this, together with the longevity of the adults, will favour iteroparity. Prolonged adult life associated with iteroparity will be enhanced by moderate levels of reproductive effort. Moreover, reproductive effort may decline with increased body size because of intrinsic negative allometry between these variables (section 6.2.3). The ability to compete for resources may be enhanced by specialised exploitation methods and this will tend to result in narrow 'niches'.

Because these properties are advantageous under conditions of population equilibrium, during which net population growth is prohibited by the density-dependent effects of competition, MacArthur and Wilson denoted them as characteristic products of '*K*-selection' (Table 6.1), referring to the symbol for the equilibrium population density, or 'carrying capacity' in the logistic growth equation. '*K*-selection' may be self-reinforcing, since large size renders organisms less vulnerable to environmental deterioration, making their populations more stable, so increasing the likelihood of chronic competition and increasing the advantage of high competitive ability (Horne, 1978). The positive feedback associated with both '*r*-' and '*K*-selection' might be expected to force organisms into either mould, leaving few intermediates.

Despite the potentially self-reinforcing nature of *r*- and *K*-selection, however, some ecologists have felt that the wide variety of ecological circumstances in nature will produce a gradation between extreme forms. If this be the case, then it is meaningful only to ascribe species to relative positions along the '*r*-*K* continuum' (Pianka, 1970). For example, *Lacuna vincta* and *L. pallidula* are small, annual, semelparous snails that seem to be near the *r*-selected end of the continuum (Grahame, 1977). But *L. vincta* is liable to undergo more violent fluctuations in population density, exerts a higher reproductive effort, produces over an order of magnitude more eggs, which are smaller and develop into planktonic veligers rather than having direct development, and is a more generalised grazer than *L. pallidula.* Therefore, *L. vincta* appears to be more strongly *r*-selected than *L. pallidula,* which may be regarded as *K*-selected relative to its congener.

When extremes of variation at each stage in the molluscan life cycle are tabulated (Table 6.2), a close parallel with the characteristics associated with *r*- and *K*-selection is obtained (cf. Table 6.1). The life-cycle data, however, are drawn up independently for the successive stages and do not necessarily represent natural sequences. Certain cells in the table may be transposed in particular cases. The simple concept of *r*- and *K*-selection, therefore, cannot explain all patterns of variation in life histories.

Some littorinids, for example, show signs of *r*-selection. They mature and remain at a small size, live for only two years or less, are semelparous and are unspecialised grazers. But instead of producing many young at a low level of parental investment, they do the opposite. *Littorina neglecta* is a small snail that lives for about 18 months and inhabits empty barnacle shells on moderately

Table 6.2: Extremes of Variation within Stages of the Gastropod Life Cycle. From Runham (in press)

Egg	Larva	Juvenile		Adult	
Simple zygote	Prolonged duration free-living	Short duration, little growth	Short life	One breeding period semelparous	Many small eggs
Complex with nutrient store	Absent or only within egg	Long duration a lot of growth	Long life	Many breeding periods, iteroparous	A few large eggs

exposed European shores (Fish and Sharp, 1985). In spite of its small size (⩽ 5mm shell height), *L. neglecta* broods embryos that are approximately as large as those borne by the larger sibling species *L. saxatilis* (Hannaford-Ellis, 1983), but because of restricted space in the brood chamber, the number of brooded embryos is much smaller in *L. neglecta.* Small size enables *L. neglecta* to shelter inside empty barnacle shells, secure from wave action, desiccation and large predators. Small size in *L. neglecta,* therefore, has probably evolved in response to the resource provided by empty barnacles, a resource that is widespread and, on a spatial scale exceeding a few metres, is relatively stable. Brooding may simply be a consequence of small size: stored energy reserves will be insufficient to produce the large numbers of eggs, albeit small ones, necessary to compensate for the heavy mortality suffered by independent embryos (section 6.5.2). On the other hand, even the production of relatively few big eggs will demand a large share of the resources of such a small animal, decreasing survivorship and favouring maximal reproductive effort and semelparity.

Territorial limpets shows clear signs of *K*-selection. They grow large, potentially live for many years, are iteroparous, are dietary specialists and their population densities tend to be kept close to the carrying capacity of the habitat by contest competition for space on the substratum (Stimson, 1970; Branch, 1975; section 7.2.2). But instead of producing few young at a high level of parental investment, they produce millions of eggs at a minimal level of parental investment per offspring. Prolific spawning is associated with the dissemination of planktonic larvae and is necessary to offset the heavy juvenile mortality accompanying this method of dispersal (section 6.6). Prolonged iteroparity insures against unfavourable environmental conditions for larvae in some years while affording the opportunity to benefit from high recruitment in good years. Widespread dispersal is advantageous because the adults are relatively immobile and chances of juvenile recruitment may be equal or greater far from the parents than near to them.

Evidently, the simplicity which gave the original concept of *r*- and *K*-selection its great heuristic value in early attempts to rationalise the observed diversity of life histories is too restrictive for complete generality. Indeed, subsequent theoretical work has shown that there may be more than one optimal evolutionary response to a selective regime. For example, the production of many small or fewer large offspring may be advantageous under conditions both of *r*- and *K*-selection (Christiansen and Fenchel, 1977). This prediction is corroborated by the previous examples of littorinids and territorial limpets and is further borne out by the life histories of mud-snails inhabiting Danish brackish-water habitats (Lassen, 1979). *Hydrobia neglecta* is an '*r*-selected', annual, semelparous species exerting a relatively high reproductive effort, having a great tendency to disperse by flotation and characterised by episodes of rapid population growth. The annual fecundity is about 300 eggs of about 164 μm in diameter, deposited singly in capsules and hatching as crawling young. *Hydrobia ulvae*, *K*-selected relative to *H. neglecta*, is a perennial, iteroparous species exerting a lower reproductive effort, having less tendency to disperse by flotation and having competitive superiority. The annual fecundity is about 3500 eggs of about 78μm in diameter, hatching as planktonic veligers.

In addition to differences among species, patterns of life history may also vary intraspecifically when populations experience different environmental regimes. Californian populations of the limpet *Collisella scabra* occupy a wide intertidal range, but because individuals home to scars on the rock, they are restricted to particular heights on the shore. Those living high on the shore experience a strongly fluctuating food supply caused by seasonal interactions between tides and climate. Consequently, the high-shore limpets grow and spawn in seasonal bursts. Recruitment is relatively low and the population is susceptible to density-independent mortality caused by desiccation and heat stress, with the result that competition for food is slight and individuals grow large (Sutherland, 1970). Those living lower on the shore experience a more stable food supply because the rocks are submerged more regularly. Growth and reproduction are more even throughout the year, but because recruitment is higher and catastrophic density-independent mortality is rare or absent, the population is maintained at a high density, causing strong intraspecific competition for food, slower growth and smaller final body size. The life-historical differences between high- and low-shore populations of *C. scabra* are phenotypic responses to environmental conditions, genetic differences being precluded by the mixing and presumably random settlement of the planktonic veligers.

Life-historical differences may also occur among latitudinally separated populations of a species, and on this spatial scale, even with planktonic larvae, there is scope for genetic differentiation. *Tegula funebralis* grows more slowly but lives longer and reaches a larger size in Oregon than in California. The northern snails spawn once and the Californian snails several times a year. Reciprocal transplantation of snails between northern and southern populations suggests that the latitudinal differences in life history have a genetic basis (Frank, 1975). Greater longevity in the northern snails may be selectively advantageous because it provides reproductive insurance against the variable success of recruitment among their populations. Repeated spawning during the year may be advantageous in the southern snails because it compensates for their shorter lifespan, itself correlated with faster growth in the warmer climate. Circumstantial support for this interpretation can be found among species of *Nucella* occupying the same geographical range, but having direct development and more even recruitment and lacking any latitudinal change in life-history characteristics (Spight, cited by Frank, 1975).

Similar latitudinal differences in life history are also present in *Notoacmea scutum* (Phillips, 1981). In Washington State the limpet grows slower, lives longer and reaches a larger size than in California where recruitment is more unpredictable. Whether or not this represents a genetic cline has yet to be determined, but the possibility seems quite likely. Whereas the northern populations exhibit the relatively '*K*-selected' features of larger size, slower growth and greater longevity (see Table 6.1), they also possess the seemingly more '*r*-selected' feature of reaching sexual maturity at a smaller size (9 mm shell length in Washington State, 16 mm in California). Because the southern snails grow faster, the difference is less in terms of the age at first sexual maturity. Nevertheless, it provides yet another example that contravenes the simple concept of '*r*- and *K*-selection'.

7 BIOLOGICAL INTERACTIONS, ECOLOGY AND ZOOGEOGRAPHY

7.1 Introduction

In performing its own functions of feeding, growth and reproduction, an organism is almost certain to influence others or to be influenced by them. The main categories of biological influence are competition, parasitism and predation. Competition can strongly influence productivity and survivorship, sometimes having important population-dynamic or evolutionary effects; parasites can render their hosts reproductively defunct, and predation may influence community structure or generate a co-evolutionary 'arms race'.

7.2 Competition

7.2.1 Introduction

Unless other factors greatly reduce recruitment and survivorship, populations in a finite habitat are bound to suffer competition for resources. The logic is so compelling, and yet unequivocal data are so difficult to obtain, that biologists have spent their working lives attempting to quantify natural competitive interactions. Interpretational difficulties arise because the effects of competition on survivorship, growth and fecundity are not easily distinguishable from those of other factors, such as the weather and habitat conditions. Carefully controlled experimentation is therefore a necessary technique of investigation. Rocky-shore gastropods, because of their accessibility, ease of manipulation and recognisable resource requirements, have proven to be useful subjects for the experimental study of competition (Underwood, 1979).

7.2.2 Intraspecific Competition

As the density of a population increases, the per capita supply of resources diminishes, eventually to a point where it begins to reduce the productivity or survivorship of individuals. If all of them gain access to some of the resource, the individuals are said to engage in 'scramble competition'; but if they hold contests in which only the winners gain the resource, they are said to engage in 'contest competition'.

Scramble is commoner than contest competition among gastropods, as it is among most invertebrates. *Nerita atramentosa* grazes intertidal micro-algae on Australian rocky shores. By caging the snails at different densities, Underwood (1976) showed that high densities reduced the growth rate of juveniles, reduced the body weight of fully grown adults and increased the adult mortality rate (Figure 7.1). Since *N. atramentosa* are mobile, non-territorial snails, food and not space must have been the critical resource. Scramble competition for food could, therefore, control their population density, operating through adult mortality rather than through recruitment. Control via recruitment is possible only when populations produce their own recruits. *N. atramentosa* have planktonic larvae and because of the associated dispersal, recruits are just as likely to originate from elsewhere as from within the population. The cage experiments show what is possible, but the frequency of population control by competition for food remains unquantified. Recruitment from planktonic larvae is characteristically so variable among places and between years that it is likely to have a strong but unpredictable effect on population density (Underwood *et al.*, 1983). Intraspecific competition will therefore probably occur sporadically, corresponding to prior incidences of heavy recruitment.

More persistent competitive control of population density might occur among gastropods with direct development, where recruitment is less variable and originates from within the population. Establishing the natural occurrence of competition and quantifying its effect on population density, however, requires intense study over many generations of the organism and has not yet been attempted adequately with marine gastropods.

Contest competition requires resources that can be defended against contestants and a behaviour pattern whereby losers retreat from the contest. Space on rocky surfaces can be patrolled and it is a source of algal food for grazers. Contest competition, in the form of territoriality, has evolved independently in certain limpets. The Californian *Lottia gigantea* (Stimson, 1970) and the South African *Patella longicosta*, *P. tabularis* and *P. cochlear* (Branch, 1981) are territorial as adults. *Lottia gigantea* defends a territory of some 900 cm^2, visiting each portion about once every 4 days while

Figure 7.1: When *Nerita atramentosa* Are Caged at Increasing Densities, Intraspecific Competition Exerts a Progressively Greater Effect on: A, Juvenile Growth Rate (Estimated from Increments in Shell Length, mm.day^{-1}); B, Adult Body Weight, g and C, Adult Mortality. After Underwood (1976)

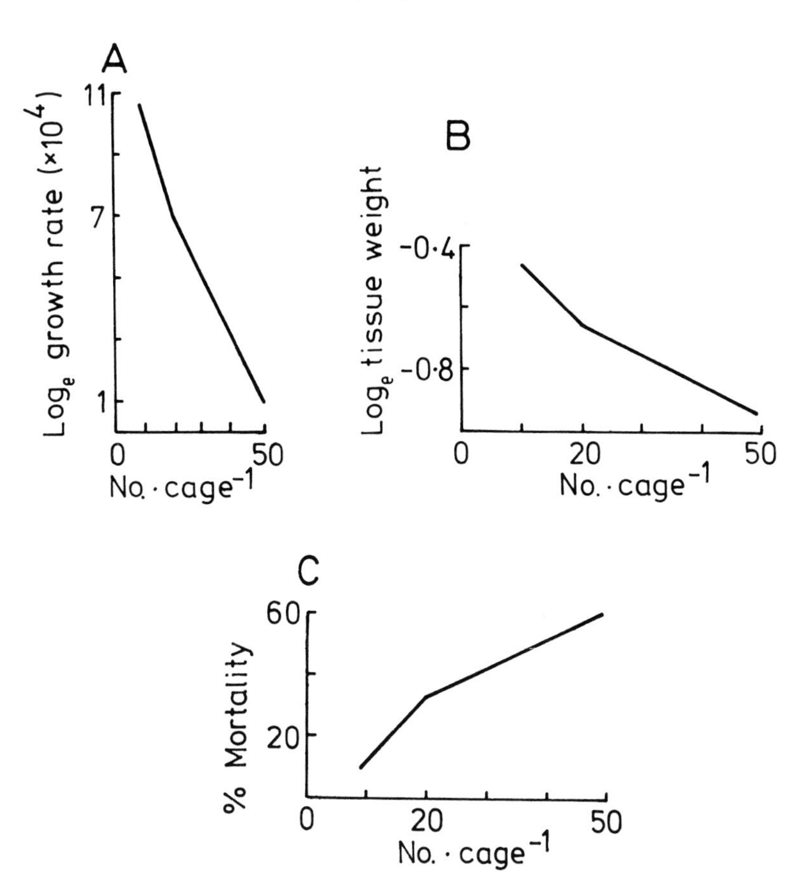

grazing the green-algal film growing within the territory. When neighbouring *L. gigantea* meet, the one nearest to its own scar is more likely to be aggressive and causes the other to recede. Removal of a territory-holder allows other herbivores to move in and overgraze the algal film within two weeks. Territoriality enables individuals to defend sufficient food for a sustained yield, which increases with territorial size as the limpet grows (Figure 7.2A). On encountering another grazer, *L. gigantea* pushes against the intruder's shell, sometimes dislodging it and causing it to be

Figure 7.2: A. Territorial Area Is Positively Correlated with Body Size, so that as *Lottia gigantea* Grows, the Amount of Algal 'Garden' Keeps Pace with the Limpet's Food Requirements. After Stimson (1970). B. The Maximum Force Exerted by *Lottia gigantea* during Territorial Contests Compared with the Force that Other Limpets Can Withstand before Being Dislodged. Moving limpets have poorer grips than stationary ones, and non-homing species have poorer grips than homers, which can withstand forces far greater than *Lottie* can exert. After Branch (1981)

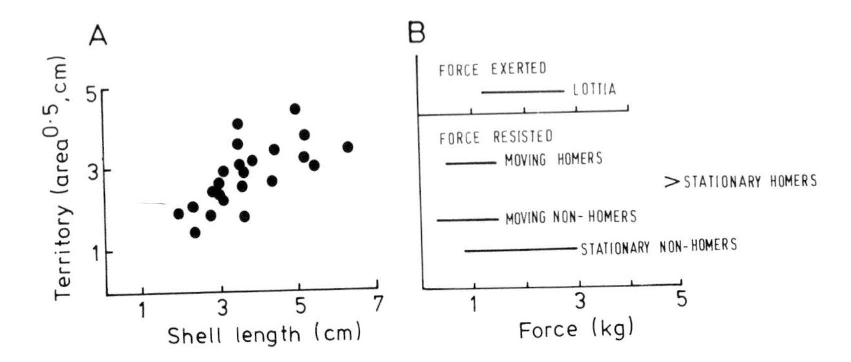

swept away in the waves. *L. gigantea* can exert sufficient force to dislodge most other limpets (Figure 7.2B), but these have a well-developed escape response that usually forestalls dislodgement.

Patella longicosta lives as a juvenile on the shells of adult limpets or of other gastropods encrusted by the red alga *Ralfsia expansa*, which is used as food. On outgrowing this substratum, juveniles move on to the rock where they feed on the calcareous encrusting alga *Lithophyllum*. During this period the juveniles lose weight and fail to mature sexually. Adults always occupy a territory containing a 'garden' of *Ralfsia*. To mature, juveniles must establish territories, probably either by finding a small patch of *Ralfsia* or by claiming a territory relinquished on the death of the owner. Territory-holders graze during high tide, and if one strays into the territory of another, the occupant reacts to contact first by extending its fringe of pallial tentacles, then within a few minutes by pushing its shell hard against the intruder, who invariably retreats.

Ralfsia within territories does not form a continuous sheet as it does elsewhere, but has a distinctive reticulated pattern (Figure 7.3A) caused by a highly organised system of grazing. If the occupant of a territory is removed, other non-territorial herbivores move in and soon eradicate the *Ralfsia* by excessively intense, random grazing. Experimental cropping of *Ralfsia* in strips increases its growth rate (Figure 7.3B), suggesting that the specialised grazing behaviour of *P. longicosta* establishes a growth pattern that maximises the sustainable yield, perhaps by increasing the surface-area to volume ratio.

Figure 7.3: A. *Patella longicosta* with its 'Garden' of *Ralfsia expansa*. The garden is about 10 cm in diameter. Compare the reticulate pattern of the *Ralfsia* with pattern c in Figure 7.3B. B. Cumulative Growth of *Ralfsia expansa*: (a) Left as a Continuous Encrustation; (b) Cut into Close Strips; (c) Cut into More Widely Spaced Strips. Treatment (c) approximates to the spacing of ridges in *Ralfsia* 'gardens' and enhances the algal growth rate. After Branch (1981)

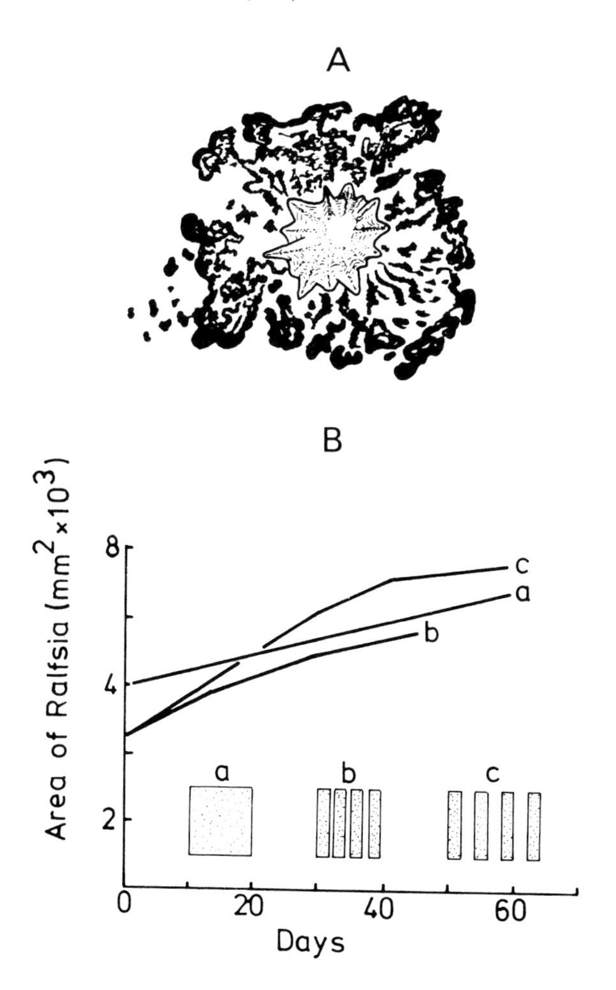

As is usual among cases of contest competition, territoriality of limpets involves an element of scramble competition, particularly marked in *Patella cochlear* which often occurs at high densities (Branch, 1975). When young, *P. cochlear* feeds on *Lithophyllum*, but adults each establish a small territory with a 'garden' of the red algae *Herposiphonia heringii* and *Gelidium micropterum*, the latter being found only in these gardens. Juveniles must establish territories between the adults, gradually increasing the size of the territories as they grow. *P. cochlear* grazes its 'garden' by revolving about the centre of the territory (Figure 7.4) and in so doing prevents the establishment of juveniles. Fighting is uncommon between adults, which do not move laterally to graze; but by excluding other grazers from their immediate vicinity they can maintain a 'garden' producing a sufficient sustained yield. At high densities, space becomes limited and restricts territorial expansion. Since the size of the algal garden governs the food supply, there is less food per individual and, since growing territory-holders are affected, scramble competition ensues. Consequently, maximum adult size and fecundity decline as population density increases (Figure 7.5).

7.2.3 Interspecific Competition

Competition from other species may depress productivity and survivorship in the same way as intraspecific competition, but with the possible additional effect of promoting 'niche separation'. Competition is likely to be most intense between closely similar species because they probably use resources in similar ways. If competition between them persists for many generations, natural selection will favour individuals of each species that use the resources in the most dissimilar ways. In different localities on the Alaskan coast, four species of *Neptunea* occur sympatrically and allopatrically. These neogastropods feed on bivalves and polychaetes, and whereas their diets overlap considerably when allopatric, there is significantly less overlap when the species are sympatric (Shimek, 1984).

Differential exploitation of resources is sometimes associated with morphological differences, which may become genetically fixed in the competing populations ('character displacement'). *Hydrobia ventrosa* and *H. ulvae* are deposit-feeders assimilating the micro-organisms among sediment particles (Figure 2.5A). Presumably for mechanical reasons, the size of ingested particles is

Figure 7.4: The Territorial Limpet *Patella cochlear* (40 mm long) Maintains a 'Garden' of Red Algae by Controlled Grazing during which the Limpet Rotates about a Central Point. As a result, the garden is only a little greater in diameter than the length of the limpet. Juveniles (bottom left) graze the surrounding calcareous encrusting alga *Lithophyllum* and do not set up territories

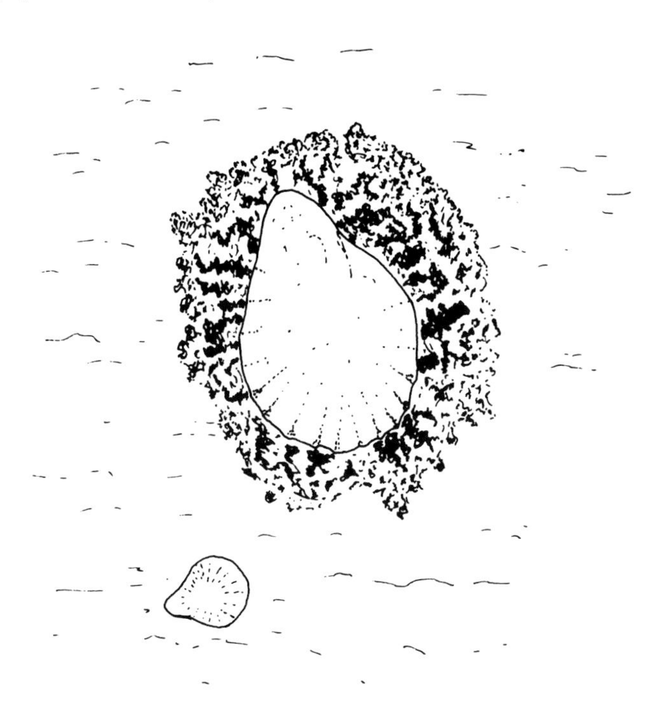

related to body size and the relationship is quantitatively similar between species (Figure 7.6A). When occurring alone, both species are of similar mean size, but in coexisting populations *H. ventrosa* tends to be smaller than *H. ulvae*, exploiting a different fraction of the sediment (Figure 7.6B). Since at the time of investigation the Limfjord, in which the populations live, was only 150 years old and the snails are annuals, the character displacement must have occurred within less than 150 generations.

A problem with interpreting the *Hydrobia* data as being character displacement is that small *H. ulvae* have to compete with numerous *H. ventrosa* of the same size before growing large enough to eat the biggest diatoms. However, character displacement could still be advantageous by making food more plentiful to adults, so increasing their reproductive output. Unfortunately, the

Figure 7.5: Reproductive Output of *Patella cochlear*, Measured as the Loss in Weight of the Gonad after Spawning, Decreases as Population Density Increases. After Branch (1981)

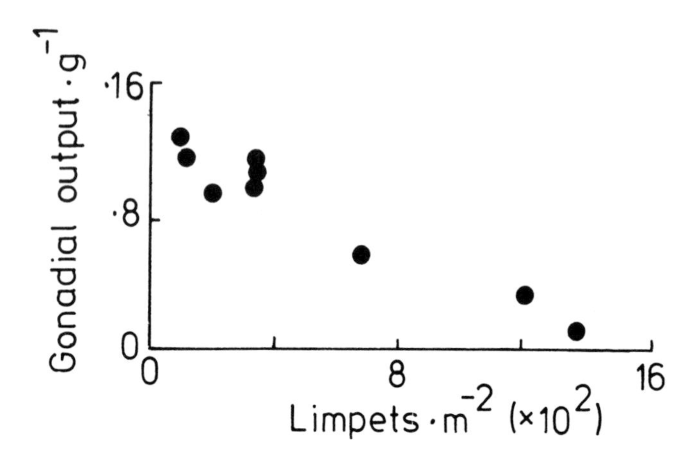

possibility cannot be ruled out that the shift in mean body sizes of *H. ventrosa* and *H. ulvae* could be due, at least in part, to environmental differences associated with allopatric and sympatric populations (Connell, 1980). Moreover, differential usage of resources (resource partitioning) may occur independently of body size because there are differences in the digestive enzymes. *Hydrobia ventrosa* has high levels of β-glucosidase and trypsin, perhaps associated with the digestion of marine fungi and bacteria; whereas *H. ulvae* is richer in α-glucosidase, suggesting the dietary importance of green algae and certain diatoms (Hylleberg, 1976).

Differential usage of resources by character displacement or by physiological mechanisms is part of the wider phenomenon of 'niche separation' among 'guilds' of potentially competing species. On comparing a variety of co-occurring animals, Hutchinson (1959) concluded that the size of the feeding apparatus must differ by a factor of about 1.3 between competitors in order for them to persist together. It is now known that there need be almost no limit to the similarity of co-occurring competitors if spatial and temporal heterogeneity of the biological and physical environment is sufficiently great. The habitats of subtidal coral reefs, however, are thought to be very stable and they support an impressively diverse assemblage of gastropods. Among these are a set of *Conus* species that feed specifically on polychaetes. With so many species

Figure 7.6: A. Species of *Hydrobia* are Deposit-feeders and the Size of Particles Ingested Is Correlated with the Size of the Snail. When they occur together in the same habitat (upper figure), *Hydrobia ventrosa* (solid line) and *H. ulvae* (dotted line) tend to eat smaller and larger particles respectively, but when in separate habitats (middle and lower figures), they eat a similar range of particle sizes. B. When they Occur in the Same Habitat (Upper Figure), *H. ventrosa* (Solid Bars) Tends to Be Smaller than *H. ulvae* (Open Bars), whereas in Separate Populations they Are More Similar in Size (Middle and Lower Figures). This difference in size-frequency structure of sympatric populations is thought to be genetically determined 'character displacement'. After Fenchel (1975)

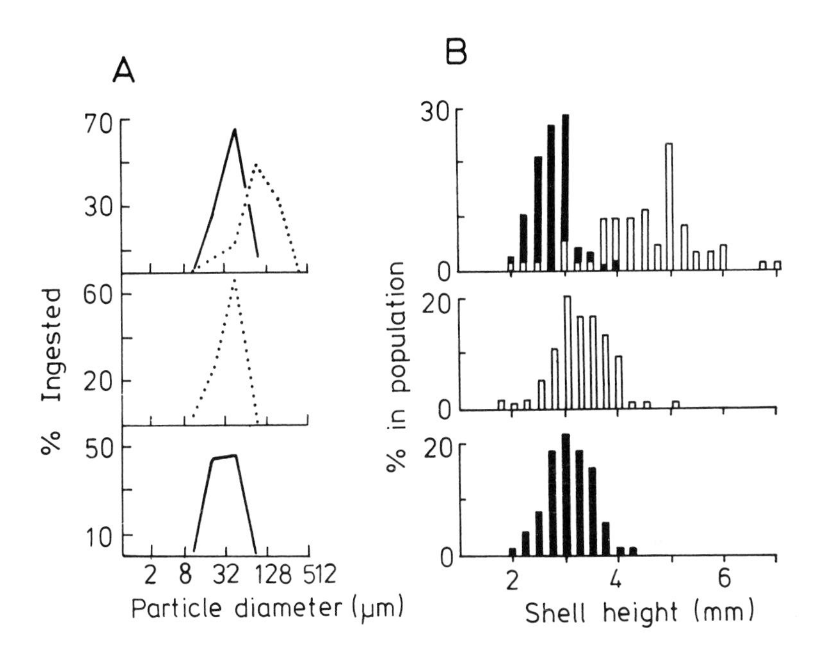

feeding on similar prey, it may be expected that they should exploit different niches in response to persistent interspecific competition.

As with most predators, the size of prey eaten increases with the consumer's body size (Figure 7.7A), and as different species of *Conus* vary in mean size, so do their prey (Figure 7.7B). There is less overlap in mean prey size among the subtidal species of *Conus* than expected by chance, consistent with the idea that interspecific divergence of body size, hence of food size, may have occurred in response to competition. Conclusive evidence for historical, unrepeatable events is of course impossible to obtain and the data are open to alternative interpretation. For example, the species of *Conus* could have evolved allopatrically and subsequently colonised the same areas, already having developed trophic differences. Whatever the cause of its origin, feeding specialisation among adult *Conus* probably does ameliorate competition. However, competition from juveniles and smaller species can reduce the food available to larger individuals by 'remote exploitation', in which small prey are consumed that would otherwise have grown to a large size (Leviten, 1978).

Dietary overlap is not significantly different from random among species of *Conus* living on intertidal rock-benches (Leviten, 1978), so feeding differences cannot entirely explain the diversity of *Conus* co-occurring on coral reefs and in associated habitats. Generally, the number of co-occurring species is proportional to the physical complexity of the habitat (Figure 7.8), suggesting that niche separation among *Conus* involves the exploitation of different micro-habitats in addition to dietary specialisation.

Physical differences in micro-habitat also account for most of the niche separation among tropical muricacean gastropods. Twenty-three species are abundant on the intertidal rocky shores of Aldabra Atoll, dividing the habitat by zonation along gradients of tidal height (Figure 7.9) and exposure to wave action and by using different types of substratum. Most of the species drill

Figure 7.7: A. As it Grows, *Conus miliaris* Eats Larger Individuals of the Polychaete *Lysidice collaris*. After Kohn (1968). B. Different Species of *Conus* Eat Sizes of prey (in this Case Polychaetes) that Are Proportional to their Own Body Size. From Leviten (1978)

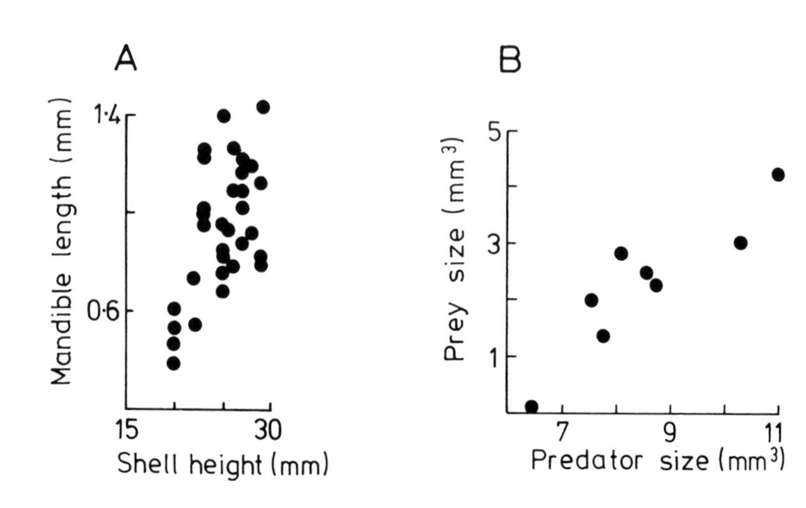

Figure 7.8: The Number of *Conus* Species in Three Habitats Plotted against an Index of Diversity ($H = \sum P_i \log_e P_i$, where P_i=Proportion of Species *i* in Sample) that Takes into Account the Relative Abundances of the Species. Both the number of species present and their compositional 'diversity' in the community are least on large stretches of bare sand (habitat I), intermediate on intertidal limestone platforms (habitat II) and greatest on subtidal coral-reef platforms (habitat III). This sequence reflects the order of structural complexity of the habitats. After Kohn (1967)

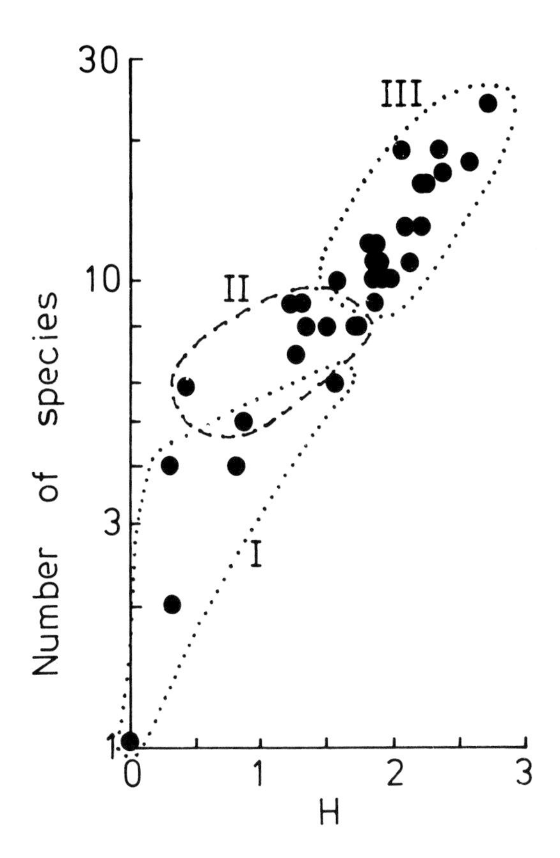

sedentary or slow-moving prey, but dietary overlap is usually low because of the different micro-habitats used for foraging.

Sometimes, there may be no scope, or insufficient time, for the evolution of niche separation when species with similar resource requirements become sympatric. In this case, one species may severely restrict the other's usage of habitat, perhaps to the extent of its exclusion. In the nineteenth century, *Ilyanassa obsoleta* was abundant on a wide variety of sheltered-shore habitats in New

Figure 7.9: Zones Occupied by Muricacean Gastropods on the Intertidal Cliffs of Aldabra Atoll, Ranging from Just Above Mean High Water Spring Tide Level to Just Below Mean Low Water Spring Tide Level. 1—*Drupella cariosa*, 2—*Morula anaxeres*, 3—*Morula marginatra*, 4—*Purpura rudolphi*, 5—*Thais aculeata*, 6—*Thais fusconigra*, 7—*Thais savignyi*, 8—*Morula granulata*, 9—*Thais armigera*, 10—*Drupa ricinus*, 11—*Drupa morum*, 12—*Drupa lobata*, 13—*Drupa rubusidaeus*, 14—*Mancinella tuberosa*, 15—*Drupella ochrostoma*, 16—*Cronia margariticola*, 17—*Nassa francolina*, 18—*Morula squamiliratum*, 19—*Morula uva*, 20—*Drupella cornus*, 21—*Coralliophila violacea*, 22—*Quoyula madreporum*. After Taylor (1976)

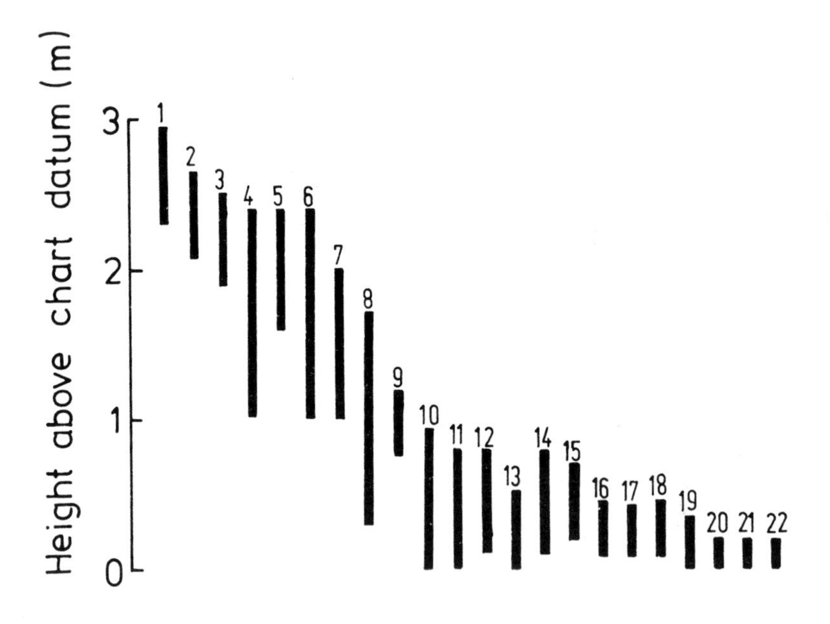

England, but after invasion by *Littorina littorea*, *I. obsoleta* became restricted to sand- and mudflats. Experiments have shown that *I. obsoleta* emigrates from areas with high densities of *L. littorea* and has thus become displaced from much of its previous range (Brenchley and Carlton, 1983). Since its introduction in the 1970s to the San Francisco Bay area, *I. obsoleta* itself has caused local exclusion of *Cerithidea californica* (Race, 1982).

7.3 Parasitism

Gastropods are primary hosts to various species of trematode (Figure 7.10) which develop in the visceral mass, deriving nourish-

Figure 7.10: Cercaria Stages of Two Trematodes that Commonly Infect *Littorina littorea*. Cercariae transmit infection from the snail to the secondary host, e.g. crabs. Left, *Renicola roscovita* (body 340 μm long); right, *Cryptocotyle lingua* (body 225 μm long). After James (1968)

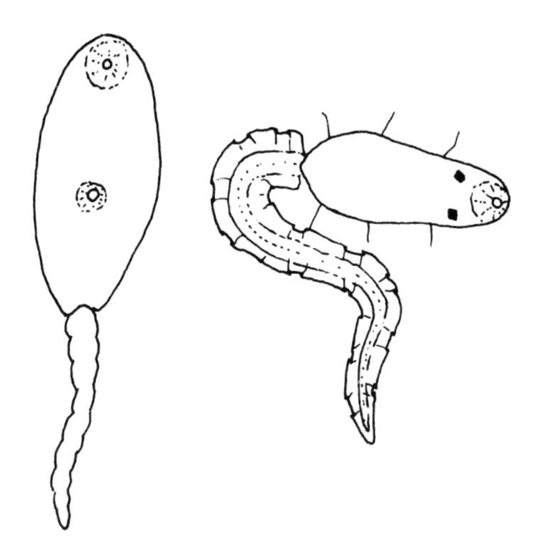

ment from the digestive gland and, in adult hosts, from the gonad. It is not known to what extent gastropods develop immunity from infection or whether it is only less vigorous individuals that become infected. Frequencies of infection seldom exceed about 30 per cent of the population and are often much lower, but increase near sources of infection such as seabird colonies where the birds (final hosts) defecate eggs on to the shore.

Some species of trematode infect only juvenile hosts whereas others infect only adults, but there may be multiple infections by several species. Frequencies of infection by different trematodes tend to vary uniquely with age of the host, some peaking earlier, some later (Figure 7.11A). The combined frequency of infection by all parasites, however, is usually a fairly smooth exponential function of the host's age (Figure 7.11B), indicating that the instantaneous risk of infection by any parasite is constant throughout the host's life and that infection is a random process.

Infection is unlikely to be the direct cause of death, since survival of the primary host until eaten by the secondary or final host is essential to the parasite. On the contrary, *Ilyanassa obsoleta* infected with *Zoogonus lasius* survive stressfully high temperatures and low salinity better than uninfected ones (Riel, 1975).

Figure 7.11: A. *Littorina littorea* Is Liable to Infection by Several Species of Trematode, Including *Cryptocotyle lingua* (Continuous Line) and *Renicola roscovita* Dotted Line). The incidence of infection as a function of host size (age) differs among trematode species. B. The Combined Incidence of Infection by All Trematodes Is an Approximately Exponential Function of Host Size, Indicating a Constant Total Risk of Infection throughout Life. After Hughes and Answer (1982)

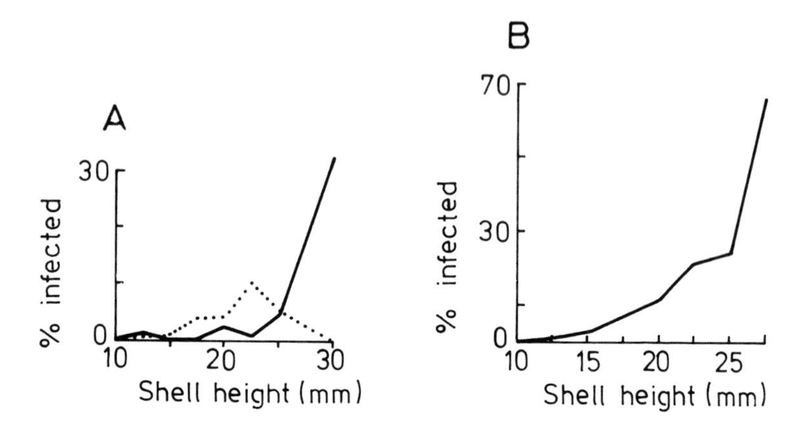

However, parasites may render their primary hosts more vulnerable to attack by later hosts. Infected *Littorina littorea* make shorter downshore winter migrations (Williams and Ellis, 1975) and so may be more vulnerable to predation. The trematode *Parorchus acanthus* affects the behaviour of *Nucella lapillus*, causing the snail to leave protective aggregations and become more vulnerable to predation by oystercatchers, the final host of the parasite (Feare, 1971b).

Trematodes multiply asexually within the visceral mass and, in adult hosts, eventually destroy the gonad. This may not necessarily jeopardise the host's life, but reduces its reproductive value to zero. The host is, in effect, reproductively dead, and if the frequency of infection were to become large within a population, the infected hosts could indirectly reduce the reproductive value of uninfected individuals by competing with them for food and perhaps by being party to unproductive copulation. *Littorina littorea* are seldom infected before their second reproductive year (Robson and Williams, 1971), but this effectively converts them from iteroparous to semelparous snails without appropriate adjustment of reproductive effort, severely reducing their lifetime fecundity (Hughes and Answer, 1984).

7.4 Predation

7.4.1 Avoidance Mechanisms

Gastropods exhibit a variety of behavioural mechanisms for avoiding predators (Feder, 1972). For these mechanisms to evolve, the predator must be a persistent source of mortality and be detectable quickly or at a distance. Most gastropods detect sudden movement by vision and withdraw into their shells. This can be demonstrated by walking at the ebbing tide among boulders used by grazing *Littorina* spp., which react by withdrawal, losing their grip and falling to the ground where they are difficult to find. Gastropods may detect invertebrate predators more effectively by olfaction or by touch. Resolution by chemoreception in gastropods (Kohn, 1961) is potentially greater than by vision, allowing them to respond to specific stimuli.

Littorina littorea crawls under rocks, into crevices or deep into macro-algal fronds when juices from crushed conspecifics are added to rock pools (Hadlock, 1980). This is an effective escape mechanism because *Carcinus maenas* takes several minutes to consume a snail, giving others the opportunity to seek shelter. Avoidance behaviour costs energy and time, perhaps increasing the risk of other sources of mortality, so is unlikely to persist in populations not exposed to the predator. *Tegula funebralis* crawls out of the water in response to the smell of crabs, predatory starfish and octopus, but individuals from populations naturally exposed to these predators respond more readily than those from elsewhere (Geller, 1982; Fawcett, 1984).

Avoidance responses should be specific, in order to prevent unnecessary reactions to animals that are not a threat. *Notoacmea scutum* normally moves upstream in a flow of water over horizontal surfaces, but moves downstream and increases its tendency to climb up vertical surfaces if the water carries the scent of the asteroid *Pisaster ochraceus* (Phillips, 1975). The limpet, however, reacts only to potential predators and not to harmless asteroids of other families (Margolin, 1964; Phillips, 1976).

Generally, gastropods react to the scent of predators by moving more quickly and in directions away from the source of danger. Members of the Strombidae, instead of gliding along, leap over the substratum by pushing and twisting with their modified foot and operculum. This is not a faster method of locomotion than gliding, but it increases agility and a single leap can place the snail out of reach of a predator. *Strombus canarium* responds to the scent of *Conus textile* by increasing the rate of leaping from about 2.7 to 6.8 cm s^{-1}, moving away from the predator (Kohn and Waters, 1966). *Nassarius luteostoma* uses the more appropriate gliding method of locomotion to move through surface layers of sand while foraging, but on contact with hunting *Polinices unifasciata* it leaps violently by extending and twisting its foot in a similar, but more uncoordinated manner than strombids (Gonor, 1965; Hughes, 1985c).

7.4.2 Neuronal Basis of Avoidance Behaviour

Tritonia diomeda, from the Pacific coast of North America, responds to contact with predatory starfish by rapid withdrawal, followed by 2 to 20 alternating dorsal and ventral flexions of the body (Figure 7.12) that cause the nudibranch to tumble randomly in the water and perhaps to be swept away from the predator by tidal currents. The cerebral, pleural and pedal ganglia of *Tritonia* are grouped into a bilaterally symmetrical 'brain' (Figure 7.13), which may be exposed through a small dorsal incision between the rhinophores and supported on a platform. After this operation, electrodes can be inserted in individual nerve cells, while the animal is free to perform most of its behavioural repertoire. Using this technique, neurophysiologists have traced a substantial part of the neuronal circuit governing the escape 'fixed-action pattern' of *T. diomeda* (reviewed in Getting, 1983).

Motoneurons causing the dorsal and ventral flexions are located in the pedal ganglia (Figure 7.13), and electrical recordings have shown that these neurons fire a single burst of action potentials in phase with corresponding dorsal or ventral flexions (Figure 7.13). Accompanying the alternating activity of the dorsal and ventral flexion neurons is the alternate firing of two other sets of neurons that may inhibit or modulate extraneous inputs from peripheral sources, but whose function has not yet been fully worked out. More importantly, a group of electrically coupled interneurons, with cell-bodies lying in the cerebro-pleural ganglia (Figure 7.13), has been identified as the 'central pattern generator', which drives the rhythmic activity of the flexion neurons. The network of the central pattern generator is comprised of synergistic and antagonistic interneurons. The synergistic interneurons reciprocally excite one another, synchronising and reinforcing their endogenous bursts of activity. The antagonistic interneurons reciprocally inhibit

Figure 7.12: *Tritonia diomeda* Responds to the Touch of a Predatory Starfish First by Withdrawing the Head and Contracting the Dorsal Gills, then by a Series of Dorsal and Ventral Flexions that Cause the Nudibranch to Tumble in the Water out of Reach of the Predator. At the end of this 'fixed-action pattern', *Tritonia* resumes normal posture and extends its gills. After Kandel (1976)

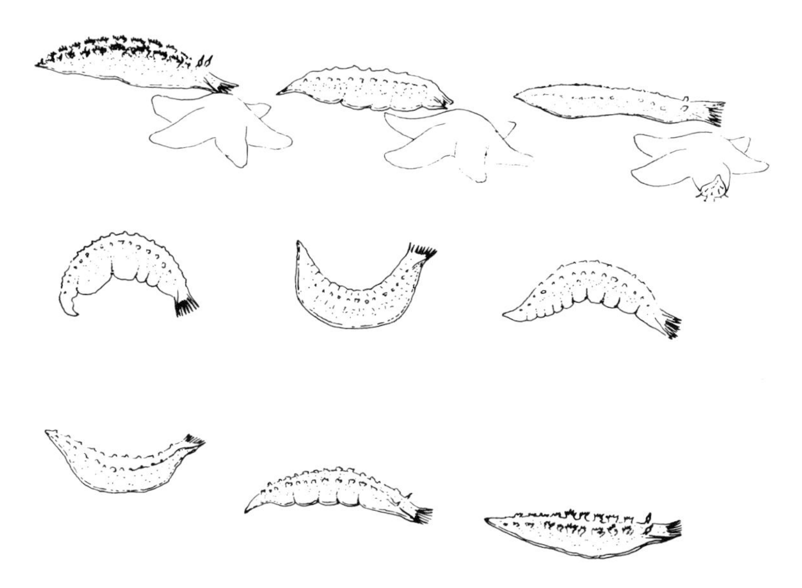

and, after a delay, excite one another. Together, the synergistic and antagonistic interneurons form an oscillator that coordinates and reinforces their endogenous activity. The oscillating central pattern generator network imposes rhythmic activity on the flexion neurons. Moreover, the 12 identified interneurons in the network are arranged in two bilaterally symmetrical subunits that are interlinked, ensuring coordinated control of left and right flexor muscles.

The escape response is initiated by the stimulation of sensory neurons innervating the anterior epithelium and whose cell-bodies lie in the pleural ganglia (Figure 7.13). Synaptic connections of the sensory neurons within the central nervous system are not well understood, but it is known that they are coupled with a bilateral pair of reflexive withdrawal interneurons, which in turn synapse with the central pattern generator interneurons. Weak stimulation of the reflexive withdrawal neurons via the sensory neurons elicits the initial withdrawal response, but not the swimming behaviour. Strong stimulation allows the reflexive withdrawal interneurons to

Figure 7.13: A. Dorsal View of the Brain of *Tritonia diomeda* Showing the Location of Neuronal Cell-bodies Involved in the Escape Fixed-action Pattern. In the right pedal ganglion the cell-bodies of flexion neurons are drawn to show the variation in shape, size and location that enables individual neurons to be identified. ce=Cerebral ganglion, dsi=dorsal interneuron, pl=pleural ganglion, pd=pedal ganglion, s=sensory neurons, tgn=trigger neurons, vsi=ventral swim interneuron. B. In the Escape Response of *Tritonia diomeda*, Dorsal and Ventral Flexions (Bottom Trace) Are Accompanied by Alternate Bursts of Firing in the Dorsal (dfn, Upper Trace) and Ventral (vfn, Middle Trace) Flexion Neurons. pd=Position detector, d=dorsal, v=ventral. After Getting (1983)

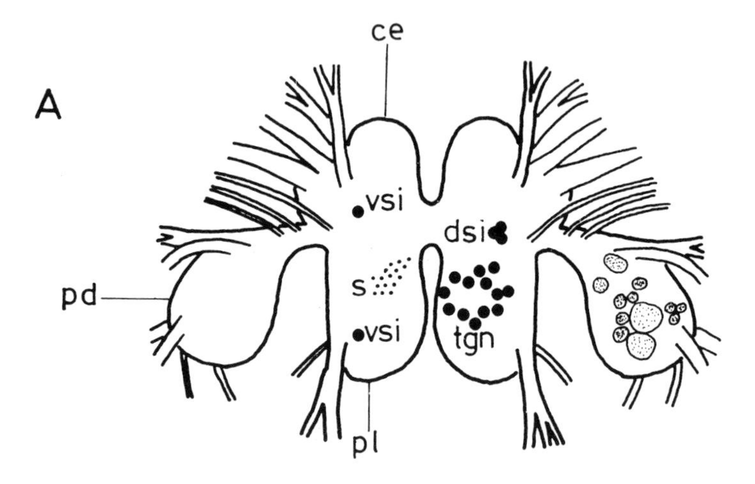

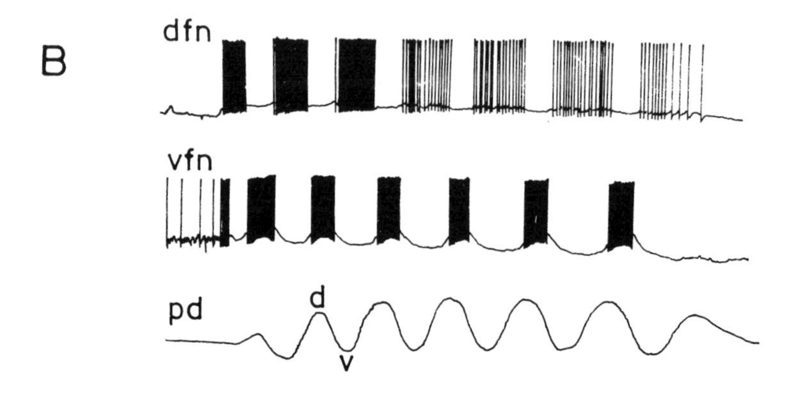

excite the central pattern generator interneurons beyond a threshold value, whereupon they inhibit the reflexive withdrawal interneuron and begin the endogenous oscillatory activity that drives the alternate firing of the dorsal and ventral flexion neurons associated with swimming.

7.4.3 Defensive Mechanisms

A minority of gastropods aggressively defend themselves when attacked. Small *Patella oculus* and *P. granatina* have a typical fleeing response to starfish and predatory gastropods, but large individuals are aggressive, lifting their shells and pulling them down hard on to the predator. Adoption of the avoidance or aggressive response depends on the relative sizes of the limpets and predators (Figure 7.14). Similarly, *Lottia gigantea* clamps the lip of its shell suddenly and forcefully on to the foot of predatory gastropods, causing them to lose their grip on the rock and be swept away by waves (Stimson, 1970). On New Zealand shores, *Melagraphia aethiops* responds to contact with the muricid *Lepsia haustrum* by lifting its shell clear of the rock and swinging it violently several times through an arc of 180°, before crawling away at two or three times the normal speed (Clark, 1958). Several species of *Conus* are capable of striking with their proboscis, and *Tonna galea* may respond to handling by ejecting sulphuric acid. But, generally, gastropods are ill-equipped for aggressive defence and must rely on other mechanisms.

Defensive secretions of mucus, sometimes containing sulphuric acid (Thompson, 1976) are widely used by gastropods lacking a shell or whose shell is covered by the mantle. Nudibranchs feeding on coelenterates incorporate nematocysts into their dorsal cerata, making them obnoxious to fish (Todd, 1981). Among prosobranchs, the shell is by far the most important defensive mechanism. Snails retreat into their shells, receding beyond reach of the predator or closing the aperture with the operculum. Limpets clamp their shells against the rock when threatened, and in homing species this may be more effective when they have returned to their scar, where the rock is worn to precisely accommodate the shell and provides a better grip because it fits closer to the substratum (Lindberg and Dwyer, 1983) and is clear of fouling organisms (Garrity and Levings, 1983). Homing to a scar may have the additional advantage of reducing desiccation. On Bermudan shores, the chloride concentration in the mantle-

Figure 7.14: The Reaction of *Patella oculus* to Predators Depends on its Relative Size. When small the limpet retreats, but when larger it pushes or clamps its shell against the predator. Smaller limpets are more offensive towards the smaller predator *Nucella dubia* (muricid gastropod) than towards the larger predator *Marthasterias glacialis* (starfish). After Branch (1981)

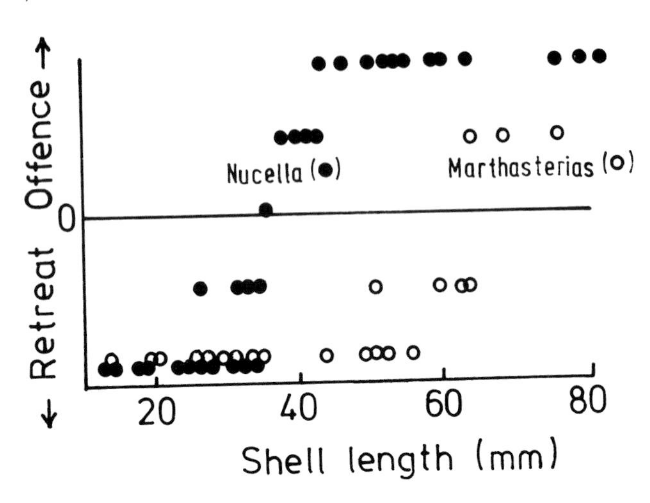

cavity fluid of *Siphonaria alternata* increases more at low tide when limpets are prevented from reaching their scars (Verderber *et al.*, 1983).

Substantial shells offer greater resistance to attack, but their secretion costs more energy (section 4.6). Also, shells of certain shapes are more resistant to predators, but may be less suitable for other functions. Consequently, shells may be expected to show strongly developed defensive properties only where the risk of predation is high. On sheltered shores where crabs are abundant, *Nucella lapillus* have thick-walled shells with narrow apertures, but on exposed shores where crabs are scarce they have flimsier shells with wider apertures (Figure 7.15A). The thicker shells are stronger (Figure 7.15B) and their narrower apertures hinder crabs trying to insert their chelae. As a result, shells of *N. lapillus* co-occurring with crabs are considerably more resistant to attack than those of individuals living elsewhere (Figure 7.15C). Resistance to crabs requires the secretion of more shell per gram of flesh (Figure 7.15D) and may constrain bodily proportions: the wider aperture of *N. lapillus* from exposed shores is associated with a wider foot that grips the substratum more tenaceously (Figures 7.15E, F).

Figure 7.15: A. Longitudinal Sections of Shells of *Nucella lapillus*. The left specimen (40 mm) is from a sheltered locality (1) where crab predation is very intense. The shell has thick walls and even though these are riddled by *Polydora* (polychaete) burrows, the shell resists crushing by the local crabs. The middle specimen is from a sheltered locality (2) where crab predation is high, but less intense. The shell has an elongate form with a narrow aperture that hinders access to crabs. The right specimen is from an exposed locality (3) lacking crabs. The shell has relatively thin walls and has a wide mouth and voluminous body whorls. After Hughes and Elner (1979). B. The Load Required to Crush *Nucella lapillus* Shells from the Sheltered Locality (1) Is Greater than that Required to Crush Shells of Similar Size from the Exposed Locality (3). After Currey and Hughes (1982). C. Below a Size of about 16 mm, All *Nucella lapillus* Can be Crushed by Adult *Carcinus maenas*, but beyond this Size, Shells from the Exposed Locality (Open Circles) Can Be Crushed More Easily than those from the Sheltered Locality (2) (Closed Circles). The 'critical size' beyond which shells are immune to crushing is greater in the sheltered than in the exposed shore forms. After Hughes and Elner (1979). D. The Mass of Shell Secreted per Unit Mass of Body Tissue Is Greater in the Stronger Shells from Sheltered Locality (1) (Closed Circles) than in the Flimsier Shells from Exposed Locality (3) (Open Circles). After Currey and Hughes (1982). E. *Nucella lapillus* with the Wider-mouthed Shells from Exposed Locality (3) Have a Relatively Larger Foot than those from Sheltered Locality (1). After Seed (1978). F. The Force Required to Detach Clinging *Nucella lapillus* from the Substratum Is Greater for the Larger-footed than for the Smaller-footed form. After Seed (1978).

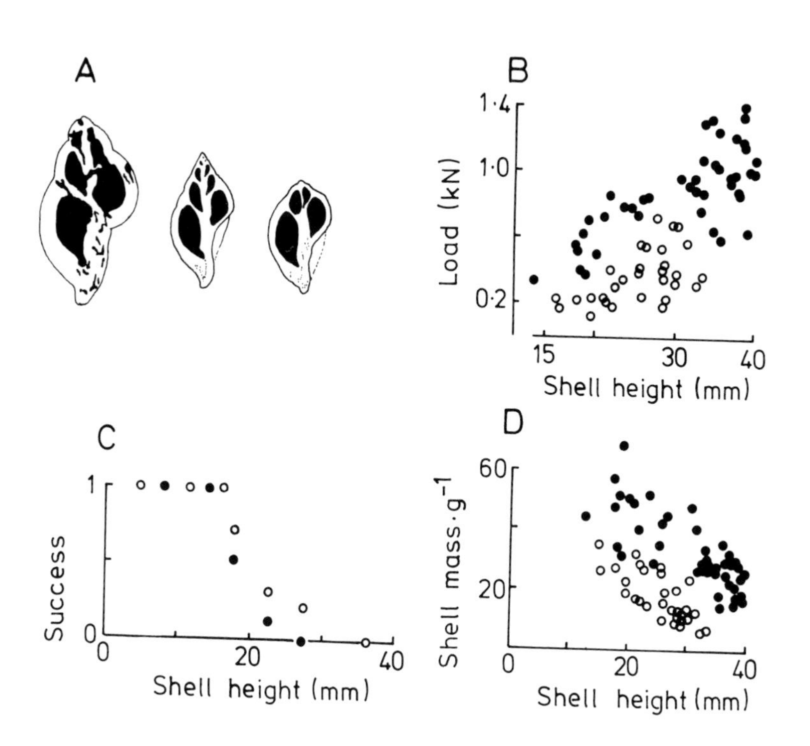

Figure 7.15 continued

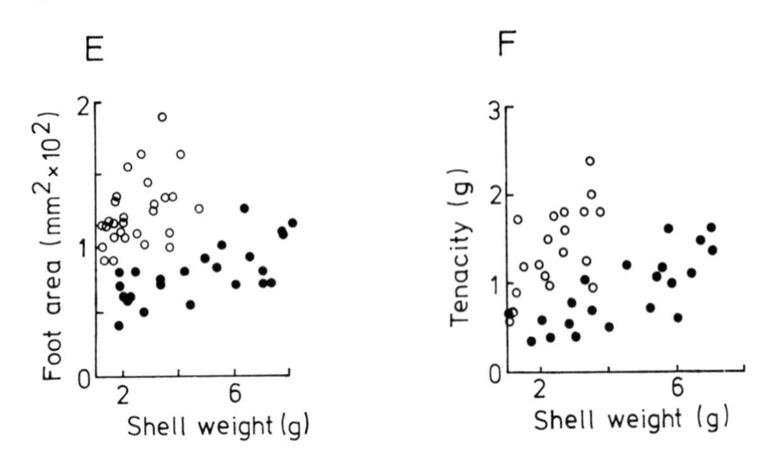

This inferential interpretation of shell morphology is corroborated by historical evidence. *Carcinus maenas* colonised the east coast of North America in the first half of the twentieth century. Museum collections of *N. lapillus* shells show that repairs to the basal whorl, damaged in a manner typical of crab attacks, rose from about 2.5 per cent in samples collected before *C. maenas* was first recorded in the various localities to about 4.4 per cent in those collected at least 7 years afterwards, and that this increase was accompanied by thickening of the shell (Vermeij, 1982).

Molluscivorous species of crab are more numerous in tropical than in temperate regions and, as might be expected, tropical gastropods tend to have thicker-walled shells with narrower apertures than temperate species (Vermeij, 1978). Interpretation of this latitudinal gradient in shell morphology is confounded by the fact that colder water impedes calcification, perhaps preventing the development of large robust shells. This physiological effect of temperature on calcification is reflected in the restriction of massive-shelled gastropods, such as species of *Cassis, Tonna, Strombus, Lambis* and *Charonia,* to tropical seas.

Tropical oceanic differences in shell morphology, however, are free from the influence of temperature and lend support to the conclusion that crab predation is an important selective force. Development of crab-resistant features is greatest among Indo–West Pacific, less among East Pacific and least among West

Atlantic rocky intertidal gastropods, paralleling a decrease in the crushing capabilities of symatric crabs (Figure 7.16). Atlantic gastropods are more easily broken by Indo–Pacific crabs than related local species (Vermeij, 1976), and the coincidental geographical patterns of shell and crab morphology strongly suggest a co-evolutionary interaction between predators and prey.

Compared with temperate species, tropical gastropods not only tend to have thicker-walled shells and narrower apertures making them more resistant to crab attack, but there is also a tendency to reinforce their shells with knobs and spines. This external sculpturing is not particularly effective against crabs, which can apply their chelae to the weaker areas between the protrusions, but is effective against wrasses and puffer-fish, which crush gastropods in their jaws and are more numerous in tropical than in temperate seas (Palmer, 1979). Knobs and spines hinder crushing by increasing the effective diameter of the shell, by localising the stress to the thicker parts of the shell (Figure 7.17) and perhaps by damaging the predator.

Gastropods themselves may exert sufficient predation pressure to cause the evolution of defensive mechanisms in their prey. Members of the subfamily Thaididae (e.g. *Nucella*) commonly feed on barnacles, using the typical muricacean method of drilling by alternate application of the accessory boring organ and radula (section 2.1.5). This is a slow process and instead of drilling completely through the exoskeletal plates of barnacles, thaidids often penetrate the sutures between the plates. Here, lateral flanges of the plates overlap, forming a wall of skeletal material similar in thickness to that elsewhere. There is, however, a thin layer of tissue between the flanges, and to reach it thaidids need to drill through only half the total thickness of the exoskeletal wall. Having reached the tissue, they inject a relaxing toxin that causes the opercular plates to gape, allowing the predator access to the prey's flesh (Palmer, 1982b).

Radiation of the Muricaceae preceded that of balanomorph barnacles during the Cretaceous (Figure 7.18A) and their predatory activities will have been a persistent force during the subsequent evolution of the barnacles. The strong evolutionary trend among balanomorph barnacles to reduce the number of exoskeletal plates (Figure 7.18B) is probably a response to such predation pressure. This would not be so if thaidids unfailingly attacked the sutures, but their drilling sites are variable and therefore a reduction in the

Figure 7.16: The Intensity of Crab Predation and Development of Structural Defences in Gastropods Are Greatest in the West Pacific, Intermediate in the East Pacific and Least in the West Atlantic. Top left, longitudinal section of *Thais melones* (East Pacific) showing thick walls and strong columella (after Palmer, 1979); top middle, *Carpilius maculatus* crushing *Strombus gibberulus* (West Pacific, after Zipser and Vermeij, 1978); top right, *Strombus gibberulus* (West Pacific) showing elongated aperture (after Palmer, 1979). The histograms show the percentages of shells collected from the East Pacific (Panama), West Atlantic (Jamaica) and West Pacific (Guam) showing features that impede crab attacks. a=Toothed apertures, b=elongate apertures, c=inflexible operculum, d=strong external sculpturing. After Vermeij (1978). Horizonal dotted lines represent an index of cross-sectional area of the chela of species of the crab *Eriphia* in the three oceanic regions. From Reynolds and Reynolds (1977)

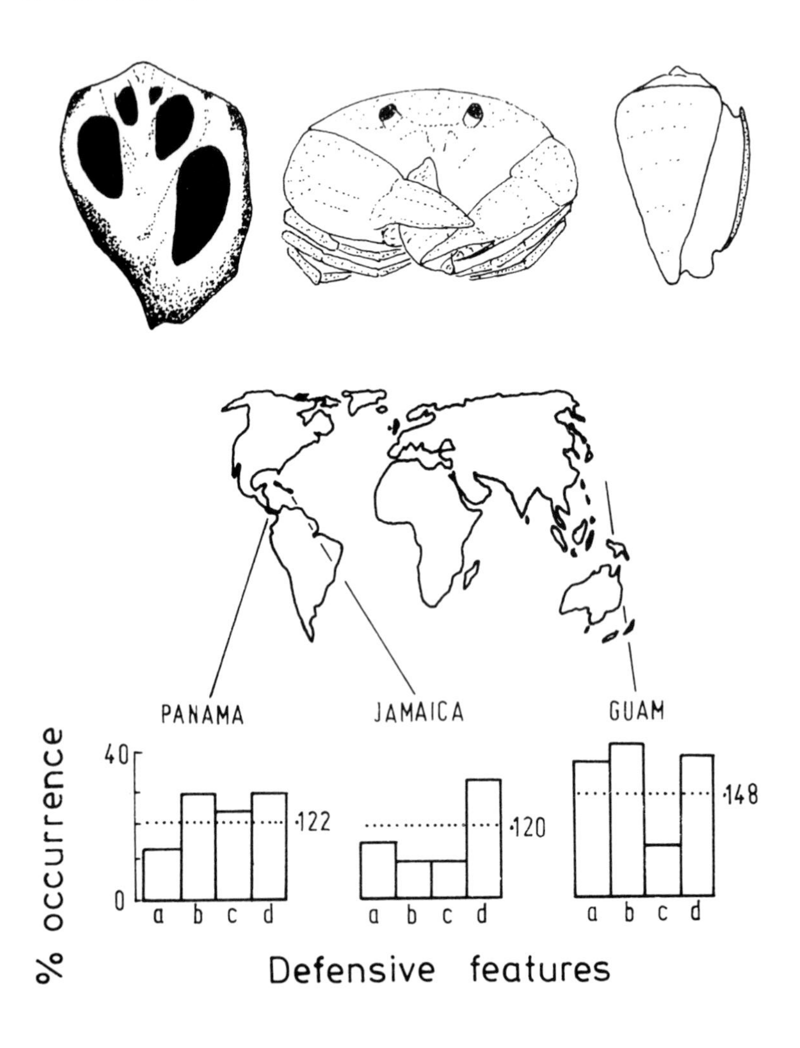

Figure 7.17: *Thais kioskiformis* Has Robust Spines that Resist Crushing by Fish. A. *Thais kioskiformis* between the jaws of the puffer-fish *Diodon hystrix*. B. Spines resist crushing by increasing the effective diameter (left) and by distributing stress over a broader area (right). C. Normal shell (left) and experimental shell with spines filed down (right). Removal of the spines renders *Thais kioskiformis* much more vulnerable to crushing by puffer-fish. After Palmer (1979).

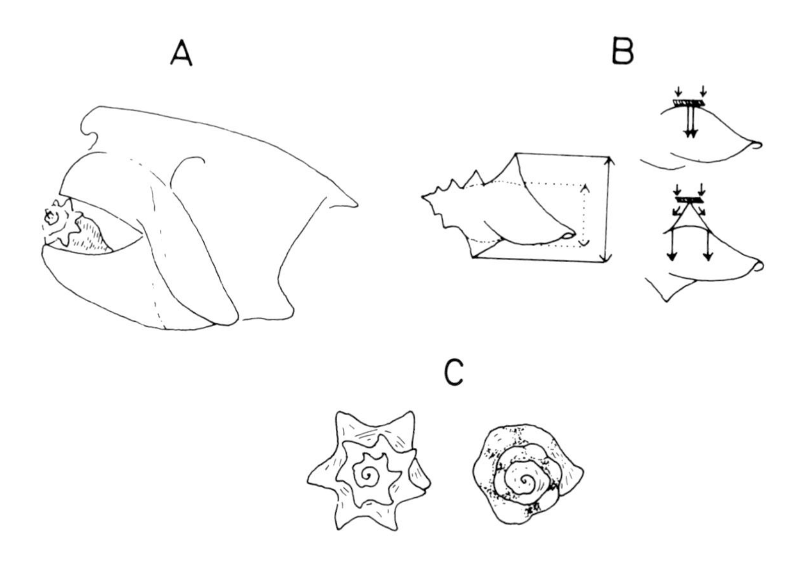

Figure 7.18: A. The Number of Genera among Balanomorph Barnacles and Muricacean Gastropods Have Increased in Parallel from the Late Cretaceous through the Cenozoic era. Bal=Balanoidea, Cht=Chthamaloidea, Cor=Corunuloidea. B. During the Radiation of Balanomorph Barnacles, the Number of Genera with 8 Skeletal Plates Has Decreased, with Corresponding Increases in the Number of Genera with 6 or 4 Skeletal Plates. After Palmer (1982b).

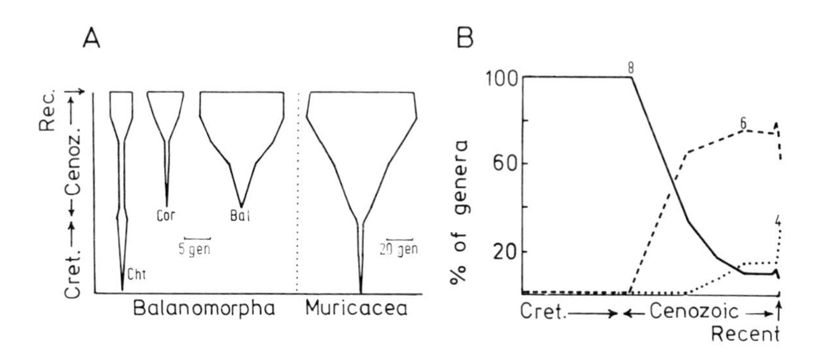

number of plates will also reduce the chance of being attacked at a suture. Lower-shore species of barnacle tend to have textural modifications of the exoskeletal surface, forming a 'thatching' that may camouflage the sutures to thaidid predators, which are commoner lower on the shore and probably detect sutures by touch, using the foot. A second evolutionary response to the risk of being drilled may be the development of numerous fine tubes within the exoskeletal walls, typical of modern balanomorphs. The tubes increase the thickness of the plates per unit of skeletal material and this may hinder application of the accessory boring organ (Palmer, 1982b).

7.4.4 Effect on Community Structure

Community structure is characterised by the identity and relative abundance of component organisms. The kinds of organisms present are ultimately determined by the physical nature of the habitat, but their relative abundances are the products of interacting physical and biological factors. Most important among the latter are competition and predation. Competitors in a community are sometimes hierarchical, so that unless interrupted by physical or biological events, the superior species will out-compete and eventually exclude the inferior ones. This situation is only likely to persist where supplies of resources are relatively stable. Such is the case on rocky shores where numerous plants and sedentary animals compete for space on the substratum; the rock surface is always there, even though its availability fluctuates with the abundance of colonising organisms. Vagaries of recruitment from planktonic spores and larvae, probably associated with variable physical conditions, can certainly have an overriding influence on rocky shore community structure. However, on certain shores, more predictable conditions exist, at least for a while, in which competition for space becomes an important structural force. In these situations, competitive exclusion of subdominants could lead to the monopolising of space by the dominants. This may be prevented locally by the interruption of competition by predation, taken in its widest ecological sense to include herbivory.

A striking demonstration of the effect of predation on community structure was made by Jones (1948) and recently by Hawkins (1981) who excluded *Patella vulgata* from areas of an exposed shore on the Isle of Man. Previously, the rocks had been relatively bare, populated by micro-algae and sporadic tufts of fucoids, but within several months of excluding the limpets, a turf of green-algal sporelings colonised the experimental strip, eventually being replaced by fucoids. In this case, intense grazing had a drastic effect on community structure, preventing macro-algae of any sort from establishing themselves except in crevices that could not be reached by the limpets.

The effect of predation on hierarchical competition greatly depends on its intensity and selectivity. Intense predation tends to eradicate species, very light predation has little effect on competition, but moderate predation may prevent competitive exclusions without eradicating any of the competitors. The latter contingency, however, depends on predators feeding preferentially on the competitive dominants; preferential predation on the subdominants will only enhance their competitive exclusion.

Both principles have been shown to operate among algae grazed by *Littorina littorea* on New England shores (Lubchenco, 1978). On these shores, rock pools vary in the number of *L. littorea* they contain and there is a parallel variation in the algal community within them. Experimental manipulation of *L. littorea* density showed that this was a causal relationship. When *L. littorea* is scarce, *Enteromorpha intestinalis* germinates along with many other ephemeral algae but soon outcompetes the others and monopolises the substratum. At high densities, *L. littorea* consumes all available edible macro-algae, including *Enteromorpha*, and the slower growing, unpalatable *Chondrus crispus* gradually forms a pure stand. At intermediate densities, *L. littorea* reduces the abundance of *Enteromorpha*, preventing competitive exclusions, and so many ephemeral and perennial algae co-occur (Figure 7.19A).

On emergent surfaces outside the rock pools, ephemeral algae grow less vigorously and *Chondrus crispus* becomes the competitive dominant at low shore levels, whereas fucoids are dominant at higher levels. When *Littorina littorea* is experimentally removed, at least 14 species of ephemeral algae germinate and live for a while in small spaces cleared by wave action, or grow epiphytically on the perennials. When at high densities, *L. littorea* consumes all the ephemerals, leaving only the unpalatable perennials, and even when at intermediate densities, it reduces the number of co-occurring algae by feeding preferentially on the competitive dominants (Figure 7.19B).

On these New England shores, *Littorina littorea* not only grazes the algae but also bulldozes sediment off the rock, with drastic

Figure 7.19: A. The Number of Algal Species Co-occurring in Tide pools on a New England Shore is Greatest when Intermediate Densities of *Littorina littorea* Graze the Algae Sufficiently to Prevent Competitive Exclusion by Dominants but Not so Heavily as to Eradicate the Dominants. B. On Rock Surfaces outside the Pools, *Littorina littorea* Feeds Mainly on Competitive Subdominants, with the Result that, Except at Very Light Densities, Grazing Reduces the Number of Co-occurring Algae. After Lubchenco (1978)

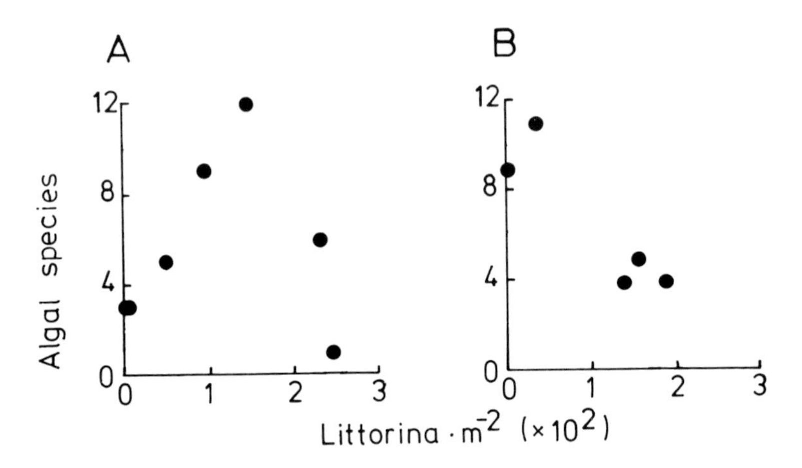

consequences to the community (Bertness, 1984). If the snails are experimentally removed, sediment begins to accumulate and an algal canopy develops which accelerates this process. Animals characteristic of soft sediment, such as tubicolous polychaetes, mud-crabs and *Ilyanassa obsoleta*, begin to predominate at the expense of those characteristic of hard substrata, such as barnacles and encrusting algae.

Grazing does not strongly influence algal associations on all rocky shores, however. At mid-shore levels in New South Wales, rock-platforms are dominated by encrusting algae, particularly *Hildenbrandia prototypus*. When the grazing gastropods *Nerita atramentosa*, *Bembicium nanum*, *Austrocochlea constricta* and *Cellana tramoserica* are excluded, sporelings of *Ulva lactuca* and other foliose algae colonise the substratum, forming a dense turf on top of the encrusting algae. Grazers therefore prevent the growth of sporelings. But when the foliose algae grow larger, they are either torn from the substratum by waves or killed by hot weather, and so physical factors, not predation, are the important determinants of the algal community structure (Underwood, 1980).

In addition to macro-algal grazers, detritus-feeders also may influence the community structure of their food organisms. At moderate grazing intensities, *Hydrobia totteni* and *Ilyanassa obsoleta* strongly depress the standing stock of filamentous blue-green algae growing on the surface of silt (Levington and Bianchi, 1981b). *Ilyanassa obsoleta* also depresses the standing stock of coccoid blue-green algae, which pass undigested through the gut of *H. totteni.* Both species of snail digest bacteria, but the division rate of these micro-organisms maintains a constant biomass in spite of the grazing pressure. How differential effects of grazing influence the coexistence of species in the microbial community is not yet understood; however, the question should be answerable by experimentation.

On rocky shores, where they are most easily studied, carnivorous gastropods exert too little predation pressure to have marked effects on community structure compared with other predators such as crabs and starfish or physical factors such as wave action. By feeding on barnacles, however, *Nucella lapillus* and other muricaceans help to generate an important habitat for numerous small organisms. After extraction of the flesh, the parietal plates of the barnacles often remain attached to the rock, forming protected cavities that can be used for shelter by animals small enough to fit inside. *Littorina neritoides,* often confined to the crevices at high levels on European shores, extends down to low levels where empty barnacles are abundant. Another small littorinid, *L. neglecta,* never grows beyond a size that can fit into empty barnacles and is largely confined to this micro-habitat, which also provides a nursery ground for the juveniles of larger-growing species such as *L. saxatilis* and *L. littorea.*

7.4.5 *Co-evolution of a Herbivore and its Prey*

Persistent predator-prey relationships cause selection for specific defences in the prey and for more effective attacking capabilities of the predator (section 7.4.3). Occasionally, the co-evolution of predator and prey has resulted in a mutual dependence, the one enhancing the survivorship and productivity of the other.

Shallow sublittoral rocks in cold-temperate and subarctic regions of eastern North America are colonised by encrusting calcareous red algae. Their morphology helps to reduce the intensity of grazing, but it also reduces productivity, making the algae slow-growing and susceptible to overgrowth and fouling. Most of the encrusting algae continually slough empty epithallial cells, but *Clathromorphum circumscriptum* retains the epithallial cells, which form a thick protective layer over the meristem (Figure 7.20A) and remain photosynthetically active. *Clathromorphum* is always associated with *Acmaea testudinalis* and experimental exclusion of the limpet causes the alga to become heavily fouled within a month by diatoms, and by blue-green and filamentous algae (Figure 7.20B). Evidently, *Clathromorphum* depends on *A. testudinalis* to keep its surface clean. To achieve this, *Clathromorphum* must provide a sustained yield of tissue to *A. testudinalis*, and calculations show that the production of epithallial cells closely matches the grazing rate of the limpet (Steneck, 1982).

Juvenile *A. testudinalis* aggregate on *Clathromorphum*, probably having been attracted to it as settling larvae. The flat, smooth surface of *Clathromorphum* affords a better grip to *A. testudinalis* than other surfaces, probably enhancing survivorship of the limpet. Even when such high-quality foods as diatoms and filamentous algae are available, *A. testudinalis* prefers to graze *Clathromorphum*. The radular teeth are aligned perpendicularly to the substratum, with blunt shovel-like tips (Figure 7.20C) that provide relatively few points of contact and so gouge forcefully into the algal crust. These features allow *A. testudinalis* to excavate the calcareous alga but impede the ingestion of diatoms and filamentous algae since the area of contact is small and gaps exist between the teeth. The co-evolutionary 'arms race' appears to have reached equilibrium because the epithallial defence mechanism of the prey has saturated the feeding capabilities of the predator and has come to support a mutually advantageous relationship between the two organisms.

7.5 Commensal Relationships

Meaning 'of the common table', commensalism in the strict sense, refers to the sharing of food by permanently associated species. Often, one species is behaviourally specialised to exploit the feeding mechanism of the other, so that the benefits of commensalism are not mutual. For example, the filter-feeding mesogastropod *Capulus ungaricus* frequently attaches itself to the shells of bivalves and to the shell of the gastropod *Turritella*

Figure 7.20: A. The Encrusting Calcareous Alga *Lithothamnion* (Left) Continually Sloughs the Epithallial Cells, so Avoiding Fouling; *Clathromorphum* (Right) Retains the Epithallial Cells, Forming a Thick Layer that Protects the Meristem from Grazers. Scale bar = 10 μm. B. *Clathromorphum* in its Natural State (Left) has Healthy Epithallial Cells Containing Plastids, and Is Free of Fouling Organisms. When *Acmaea testudinalis* is excluded, epiphytes foul the surface of *Clathromorphum* (right) and cause deterioration of the epithallial cells. Artificially heavy grazing by *A. testudinalis* crops the epithallus close to the meristem (centre). Scale bar = 5 mm. C. When in Use, the Radular Teeth of *Acmaea testudinalis* Stand at Right Angles to the Radular Ribbon. They act as shovels, scraping across the epithallus of *Clathromorphum* to produce the pattern depicted at the right. Scale bar = 100 μm. After Steneck (1982)

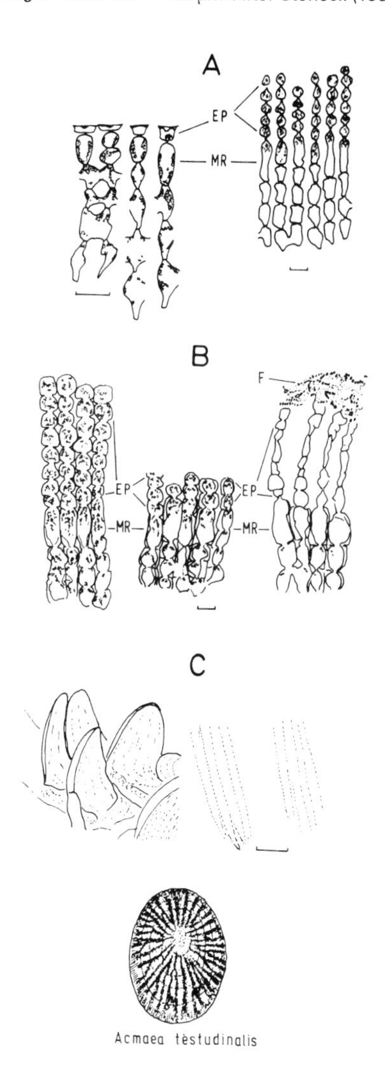

communis, occupying a position that enables it to intercept the host's inhalant current and perhaps to extend its long proboscis and steal some of the host's food-laden mucus (Fretter and Graham, 1962).

In a broader sense, commensalism can include any association that does not involve harmful effects of one participant on the other. The Australian limpet *Patelloida mufria* lives both subtidally and intertidally. Subtidal individuals occur on rock, but intertidal ones are nearly always confined to the shells of other gastropods, particularly the trochid *Austrocochlea constricta*. Experiments show that *P. mufria* greatly prefers *A. constricta* to other species such as *Cellana tramoserica*, and that it prefers mobile *A. constricta* to immobile ones. *P. mufria* attaches itself beneath the basal whorl of *A. constricta*, and the tendency of this host (unlike *C. tramoserica*), to enter rock pools at low tide, or when on emergent surfaces to trap seawater under its shell, protects the limpet from desiccation. When experimentally placed on rock, *P. mufria* quickly dry out and die. The host's movements also protect *P. mufria* from the predatory gastropod *Morula marginalba* (Mapstone *et al.*, 1984).

Although perhaps not a critical feature of the commensalism, *P. mufria* prefers *A. constricta* that have not recently been occupied by another limpet. Evidently, *P. mufria* grazes the algae growing on the host's shell and the food supply will be greater on previously unoccupied shells.

Food supply certainly is a critical factor, however, in the commensalism between juvenile *Patella longicosta* and other gastropods such as *Oxystele sinensis* on South African shores. Here, the hosts provide the only source of the specific food, *Ralfsia expansa*, that is available to the young limpets (Branch, 1975). *Ralfsia expansa* growing on rock is defended by territorial adult *P. longicosta*, and juveniles confine their grazing activities to the shells of their hosts (section 7.2.2).

7.6 Zoogeography

Patterns in the diversity and taxonomic distribution of animals about the world arise from a combination of past and present geographical and climatological conditions, together with the biological consequences of colonisation, competition, predation, extinction and speciation (Valentine, 1973; Briggs, 1974; Vermeij, 1978).

No animal is tolerant of all environmental conditions, and therefore geographical changes in climate or habitat will restrict the distribution of species, though some will have wider tolerances and hence wider geographical ranges than others. *Littorina neritoides* has its centre of distribution on the northern shores of the Mediterranean Sea, but whereas it maintains abundant, viable populations throughout most of the British Isles, it peters out in Norway (Hughes and Roberts, 1980b). There are apparently no outstandingly effective competitors or predators of *L. neritoides* in these latter areas, and the northern limits of its distribution seem to be set by a temperature regime too cool for successful reproduction. Similarly, *Cassis madagascariensis*, despite its name, is principally a Caribbean species, extending to the offshore waters of North Carolina where the warm, northbound Florida Current maintains conditions favourable for its growth and reproduction. The sandy habitat and the echinoid prey of *C. madagascariensis* extend much further north to the Cape Cod region, but the south-flowing Labrador Current is too cold for this species to survive, not least because it hinders calcification of the massive shell (section 7.4.3). Conversely, a smaller cassid, *Galeodea echinophora*, inhabits the cool, deep waters of the Mediterranean Sea and adjacent East Atlantic region, but is unable to survive warmer conditions (Hughes, 1985a) and so is excluded from shallow warm-temperate or tropical habitats, where it is replaced by species of *Phalium*.

Circumstantial evidence such as this, however, must be interpreted with caution since the possibility that competition or predation modify distributional limits cannot be discounted without appropriate data. For example, on the west coast of North America the southern limit of *Littorina sitkana* occurs at about 43°N and individuals transplanted further south succumb not only to desiccation but also to predation by crabs, which are more abundant in the high intertidal zone to the south (Behrens-Yamada, 1977).

Species extend their geographical range by dispersal, and in marine gastropods this is by crawling or drifting. Dispersal by crawling is limited by the substratum, rock-dwellers being unable to traverse large expanses of fine sediment either because they are morphologically incapable of gliding over the surface, as for example are limpets, or because their pallial system becomes choked with silt. Conversely, species inhabiting sediments may be incapable of adhering firmly to rock and easily become dislodged by currents or waves. Tropical shallow benthic species cannot traverse the deep ocean floor between islands and land masses because at these depths suitable habitat and food are absent and it is too cold. For these reasons, long-distance dispersal of gastropods is predominantly by drifting, usually of veliger larvae (section 6.6). It is generally assumed that species with direct development disperse by crawling, but some may be capable of drifting. For example, the newly hatched young of certain vermetids (Hughes, 1978) and of *Galeodea echinophora* (Hughes, 1985a) secrete mucous strings that serve as drogues, carrying the tiny snails along in water currents.

The success of dispersal by drifting relies upon favourable patterns of water circulation. Potentially suitable habitats may be denied to certain species because contrary oceanic currents prevent the spread of planktonic larvae to these areas. *Littorina planaxis* does not extend beyond latitude 43°N on the west coast of North America even though snails transplanted further north survive well and reproduce. Evidently the potential northerly range of *L. planaxis* is curtailed by south-flowing currents that prevent the northward spread of the pelagic larvae (Behrens-Yamada, 1977).

A similar mechanism may account for the scarcity of Indo-Pacific species on the shores of western Central America. Gastropods with teleplanic larvae (section 6.6) are certainly capable of surviving the long journey from the nearest source regions (Marquesas and Line Islands) in the Central Pacific, and some 45 species of gastropod have colonised islands 500-100 km offshore from Central America (Zinsmeister and Emerson, 1979). Larvae of these species could have travelled in the North Equatorial Counter-current, which flows eastwards at variable speeds of up to 2 km h^{-1}, but are also likely to have been brought across in the Cromwell Current, which forms a consistent shaft of warm east-flowing water at about 100 m depth, moving at speeds of up to 5 km h^{-1}. A similar subsurface current transports larvae eastwards across the Atlantic (Scheltema, 1968). These currents provide excellent dispersal corridors, but the relatively small number of species that have survived the journey reflects the vast expanses of open water between source regions and the nearest land. Only species with exceptionally long-lived planktonic (teleplanic) larvae can make the crossing. The vast majority of larvae that have survived crossing the 'East Pacific Barrier', however, stop short about 500-100 km off Central America, where the warm east-bound currents abate and are met by a westward flow of cold surface water (Figure 7.21). Consequently, very few of the 45 Central Pacific species of gastropod that have colonised offshore islands such as Clipperton, Cocos and the Galapagos archipelago, have also reached the mainland coast.

Gastropods might occasionally disperse by attachment to other moving objects, either floating debris carried by the ocean currents or ships and their cargo. Accidental transportation by ships is rapid

Figure 7.21: The Warm East-flowing North Equatorial Countercurrent (NECC) and Subsurface Cromwell Current (CRC) Provide Dispersal Corridors for Teleplanic (Long-lived) Veliger Larvae of Gastropods inhabiting Shallow Waters around the Line Islands and Marquesas. Such larvae colonise offshore islands, e.g. Clipperton, Cocos, and the Galapagos archipelago, but most are prevented from reaching the American mainland by cold, west-flowing currents (dotted arrows) derived from the California Current to the North and the Peru Current to the South. After Zinsmeister and Emerson (1979)

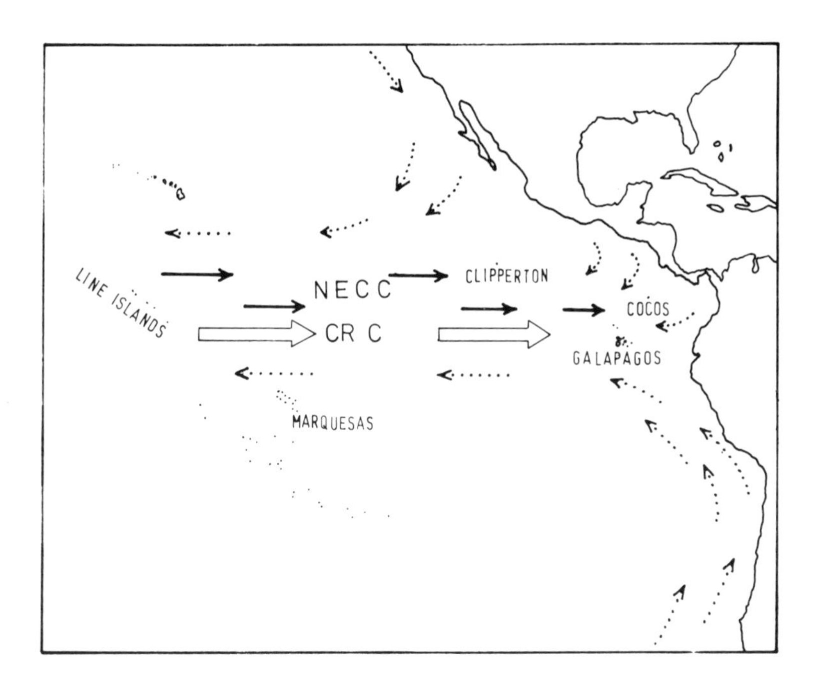

and independent of currents and may account for the presence of *Littorina saxatilis* in two estuarine ports in South Africa, separated by thousands of kilometres of tropical coastline from the nearest source populations on the African coast of the Mediterranean Sea (Hughes, 1979b).

When species colonise a new area, they might join a community lacking other species with similar ecological requirements, in which case the immigrants will, through the process of intraspecific competition, gradually expand their range of resource usage (competitive release). For example, *Conus miliaris* is the only member of its genus, apart from a sparse population of *C. sponsalis*, that has successfully colonised the geographically isolated shallow benthic habitats of Easter Island. Here, *C. miliaris* exploits 16 species of polychaete worms compared with 2-8 species where it occurs with other congeners in the central Indo-West Pacific (Kohn, 1978).

Immigrants might also join a community already containing species with similar ecological requirements. If the immigrant is competitively superior, it will displace the native species, as *Littorina littorea* displaced *Ilyanassa obsoleta* from rocky habitats in New England, and as *I. obsoleta* displaced *Cerithidea californica* from muddy creeks in San Francisco Bay (section 7.2.3). If the immigrant is competitively inferior, it will either by confined by interspecific competition to a narrow segment of its potential ecological range (the opposite of competitive release) or may become extinct in that area.

Immigrants of a species are likely to be so few that they represent only a small portion of the source gene pool, so that through the statistical effects of the 'founder principle' and 'genetic drift', their descendant populations will come to differ genetically from the ancestral one. Any such genetic divergence will be reinforced by the almost inevitable differences between selective forces operating in the source and colonised environments and by the lack of gene flow between immigrant and source populations. These factors probably explain the substantial genetic differences between European and North American populations of *Littorina littorea* that have been revealed by the electrophoresis of allozymes (Berger, 1977). *L. littorea* has been found near Halifax, Nova Scotia, in Indian shell middens at least 700 years old, but it is only during the present century that this species has spread south to the shores of New England. What caused this range extension is uncertain, but its success has been enormous and *L. littorea* is now a dominant organism on these shores. Compared with individuals from European populations, *L. littorea* from Cape Cod shows a low level of heterozygosity, probably caused by the founder effect. Evidently the North American population of *L. littorea* originated from a small sample of transatlantic immigrants and, until quite recently, was unable to expand substantially. Even though the North American population is now large, its gene pool reflects the low heterozygosity inherited from the few founder members. Increasing levels of heterozygosity will probably arise as the North American population evolves.

Although, by preventing genetic mixing, geographical isolation enhances evolutionary changes among populations, it is not necessarily a prerequisite for speciation. Species with direct development do not experience the amount of genetic mixing associated with planktonic dispersal, and small differences in their behaviour or habitat usage can effectively reduce gene flow between neighbouring populations (Janson and Ward, 1984). Consequently, if selection is sufficiently strong and persistent, such populations may undergo 'sympatric' speciation. British shores support at least three 'sibling' species of *Littorina* (Figure 7.22) that have probably evolved sympatrically from an ancestral lineage close to *L. saxatilis*. All have direct development and slightly different but overlapping habitat requirements. *L. saxatilis* is ovoviviparous (broods young; section 6.5.2) and occupies a wide range of shores from exposed cliffs to saltmarshes at mid to high spring tide level (section 6.2.3). *L. neglecta* is also ovoviviparous but is a small, short-lived species almost entirely restricted on British shores to the crevices among empty barnacles (section 6.8). *L. nigrolineata* is oviparous (lays eggs) and is confined to rocky shores exposed to at least moderate wave action, living at low to high neap tide level. Genetic differentiation among these sibling species has been confirmed by the electrophoresis of allozymes (Wilkins and O'Regan, 1980) and by breeding experiments (Warwick, 1982).

Opportunities for speciation and the coexistence of species depend on the diversity of resources available and, in the case of food, on the stability of the resources. Diverse resources provide more opportunity for species to use them in different ways (resource partitioning), so avoiding competitive exclusion (7.2.3). Potentially competing species thereby come to use a narrower

Figure 7.22: The Sibling Species *Littorina saxatilis* (Top Left, 15 mm Shell Height), *L. nigrolineata* (Top Right, 15mm) and *L. neglecta* (Bottom, 5 mm) Probably Evolved in the Absence of Geographical Isolation, Being an Example of Sympatric Speciation

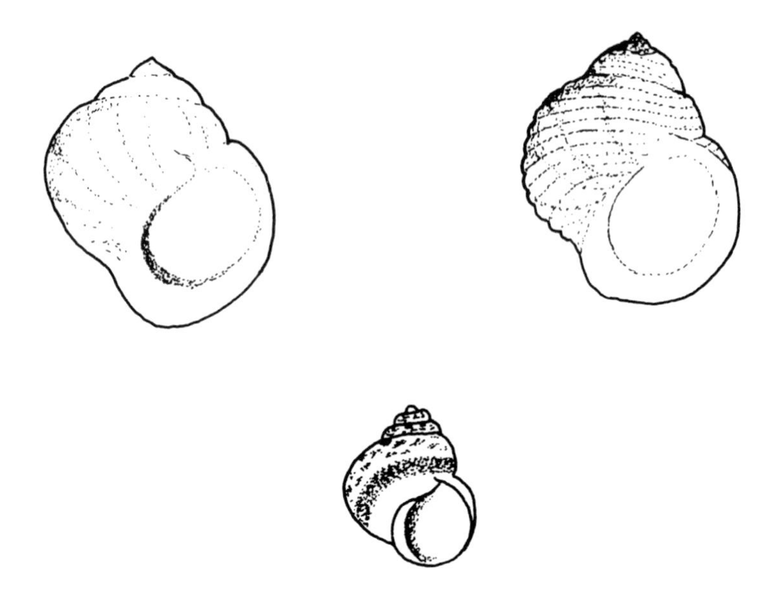

range of resources, but such specialisation in only feasible if the resources are relatively stable. For example, a specialised feeder cannot persist if its food disappears for part of the year.

This may explain why the diversity of gastropods decreases from equatorial to polar regions. Among the gastropods of the eastern Atlantic, for example, there is a sharp drop in the number of species round about latitude 40°N (Figure 7.23A), coinciding with a change from a tropical oceanic regime in which primary production is continuous throughout the year, to a temperate one in which primary production fluctuates seasonally. The changing regime of primary production evidently has a great influence on the prey of gastropods, for it is the attenuation of most tropical families of predators that accounts for the sharp drop in total number of species at the transition from tropical to temperate regions. Stable primary production in the tropics allows trophic specialisation and hence diversification of prey species, which in turn allows trophic diversification of their predators. Most tropical predatory gastropods specialise on a narrow range of prey, whereas most predatory gastropods from higher latitudes are opportunistic feeders on a wide variety of prey (Taylor and Taylor, 1977). The Turridae are exceptional, living at high latitudes yet being specialist feeders on polychaetes, as is *Nucella lapillus*, which feeds specifically on barnacles and mussels. The prey of these gastropods, however, are deposit-feeders or long-lived filter-feeders whose populations are little affected by seasonal changes in primary production.

In contrast to predatory species, the number of grazing species rises to a sharp peak between latitudes 40 and 60°N (Figure 7.23B), but again the abundance of food resources seems to be responsible for the latitudinal trend. Temperate shores are ideal for the growth of attached algae, as neither are the winters too cold nor the summers too hot. In polar regions, winter ice abrades the substratum in shallow water, and on tropical shores, the intertidal zone dries out quickly and receives such intense insolation that algal growth is limited.

The principles mentioned above can be used to explain the large differences in the faunistic diversity of the major oceanic regions.

Figure 7.23: A. The Total Species Diversity of Gastropods in the Eastern North Atlantic Drops Sharply around Latitude 40°N, where a Regime of Tropical Stable Primary Production Gives Way to One of Seasonal Primary Production. The change in species diversity is largely due to a reduction in predatory species of prosobranchs outside the tropics, perhaps associated with less predictable prey populations that are influenced by the seasonal primary production. B. The Species Diversity of Non-predatory (Largely Grazing) Gastropods Peaks between Latitudes 40 and 60°N. Here the climate favours the growth of attached algae more than in tropical or polar regions. After Taylor and Taylor (1977)

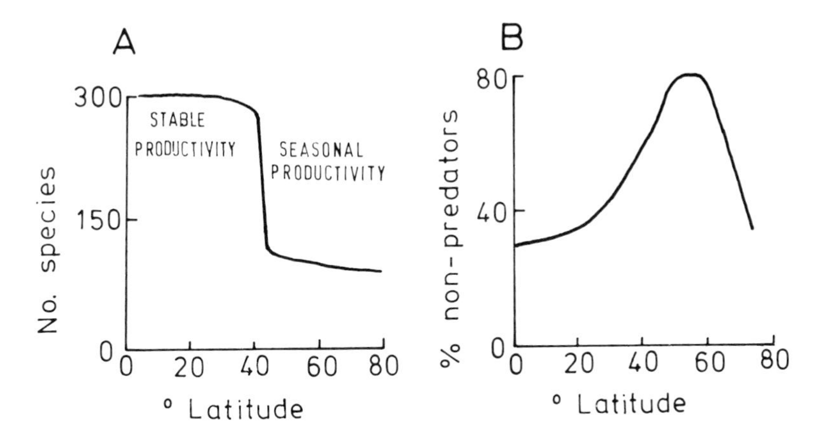

The highest diversity of shallow benthic species occurs in the tropical Indo-Pacific, followed by the tropical Pacific coast of America, the Caribbean and finally the tropical East Atlantic. Within the vast Indo-Pacific, the highest diversity is centred on the Indo-Malayan region where many factors promoting species diversity occur together. The equatorial location is associated with stable, non-seasonal primary production. The climatic stability is enhanced by the fact that the numerous archipelagos and continental land masses are surrounded by a large ocean, which acts as a heat sink, damping fluctuations in temperature. The continental shelf is dissected by numerous deep basins between islands and continents, hindering dispersal and promoting the genetic isolation of populations, whereas the extensive coral reefs in this region provide a spatially heterogeneous environment with a stable, high productivity, offering many opportunities for resource partitioning among species. Other oceanic regions are climatically less stable because they have greater ratios of continuous land surface to sea surface, with the result that the ocean becomes less effective as a heat sink. This ratio is least among the archipelagos and peninsulas of the Indo-Malayan region, greater along the West American shelf where large continents border a large ocean, and greatest in the Atlantic where large continents border a small ocean.

Other factors also influence regional species diversity. For example, the Pacific coast of tropical America would be richer if the cold inshore currents did not destabilise the temperature regime and did not restrict immigration from the Central Pacific (discussed above). The west coast of Africa is depauperate in species partly because the cold north-flowing water from the Benguela Current destabilises the temperature regime, making it unsuitable for certain tropical animals such as corals and associated fauna, and partly because there are long stretches of coastline devoid of the rocky substrata necessary for many species of gastropod.

Although oceanic differences in species diversity can ultimately be explained in terms of physical factors, biological interactions may also play an important part. In both the tropical Atlantic and Pacific, shallow benthic species diversity is greater on the western than on the eastern side of the ocean. This undoubtedly reflects the geographical differences between opposite sides of the oceans and the predominantly westward flow of warm currents suitable as dispersal corridors (Scheltema, 1968; Zinsmeister and Emerson, 1979). The imbalance might be reinforced, however, if the chances of successful colonisation by immigrants to the western ocean communities were less than for immigrants to the east (Briggs, 1974). Such would be the case if the diverse western-ocean species were more specialised and hence competitively superior in their own micro-habitats to the eastern species. Circumstantial evidence suggests that species from the diverse Indo-Pacific region are indeed more specialised than those elsewhere, but their relative competitive ability could only be ascertained by monitoring the populations of reciprocally transplanted species. An experiment of this sort probably would occur by default if a sea-level canal were to be made across the isthmus of Panama. The competition hypothesis predicts that immigrants from the more diverse, presumably more specialised and competitively superior Pacific fauna would displace native species in the Caribbean (Briggs, 1974; Vermeij, 1978).

APPENDIX: CLASSIFICATION OF GASTROPODS

This appendix includes only families and genera referred to in the text.

Class GASTROPODA

Subclass **Prosobranchia**

Order Archaeogastropoda

- *Superfamily* Pleurotomariacea
  - *Family* Pleurotomariidae: *Pleurotomaria*
  - *Family* Haliotidae: *Haliotis*
- *Superfamily* Fissurellacea
  - *Family* Fissurellidae: *Fissurella*
- *Superfamily* Patallacea
  - *Family* Acmaeidae: *Acmaea, Collisella, Lottia, Notoacmea, Patelloida*
  - *Family* Patellidea: *Cellana, Helcion, Nacella, Patella*
- *Superfamily* Trochacea
  - *Family* Trochidae: *Austrocochlea, Calliostoma, Melagraphia, Monodonta, Oxystele, Tegula, Trochus, Umbonium*
  - *Family* Turbinidae: *Subninella, Turbo*
- *Superfamily* Neritacea
  - *Family* Neritidae: *Nerita, Neritina*

Order Mesogastropoda

- *Superfamily* Littorinacea
  - *Family* Littorinidae: *Bembicium, Littorina, Nodilittorina*
  - *Family* Lacunidae: *Lacuna*

Superfamily Rissoacea

- *Family* Barleeidae: *Barleeia*
- *Family* Rissoidae: *Alvania, Rissoa*
- *Family* Hydrobiidae: *Hydrobia, Potamopyrgus*

- *Superfamily* Cerithiacea
  - *Family* Turritellidae: *Turritella*
  - *Family* Vermetidae: *Dendropoma, Serpulorbis, Tripsycha*

Family Caecidae: *Caecum*
Family Cerithiidae: *Cerithium, Cerithidea*
Superfamily Strombacea
Family Strombidae: *Lambis, Strombus*
Family Aporrhaidae: *Aporrhais*
Superfamily Calyptraeacea
Family Capulidae: *Capulus*
Family Crepidulidae: *Crepidula*
Superfamily Cypraeacea
Family Cypraeidae
Family Eratoidae: *Erato*
Superfamily Naticacea
Family Naticidae: *Natica, Polinices*
Superfamily Scalacea
Family Scalidae
Family Ianthinidae
Superfamily Eulimacea
Family Eulimidae: *Melanella*
Family Aclididae: *Cima*
Family Entoconchidae: *Enteroxenos, Stilifer*
Superfamily Heteropoda
Family Carinariidae: *Cardiapoda, Carinaria*
Superfamily Tonnacea
Family Tonnidae: *Tonna*
Family Cassidae: *Cassis, Cypraecassis, Galeodea, Phalium*
Family Cymatiidae: *Charonia*
Family Bursidae: *Bursa*

Order Neogastropoda
**Superfamily* Muricacea
Family Muricidae: *Acanthina, Cronia, Dicathais, Drupa, Drupella, Lepsia, Mancinella, Murex, Nassa, Nucella, Ocenebra, Searlesia, Thais, Urosalpinx*
Family Coralliophilidae: *Coralliophila, Quoyula*
Superfamily Buccinacea
Family Buccinidae: *Buccinum, Neptunea*

---

* *For a different interpretation of neogastropod superfamilies see Ponder (1973).*

Family Nassariidae: *Bullia, Ilyanassa, Nassarius*
Family Fasciolariidae: *Fasciolaria*
Superfamily Volutacea
Family Olividae: *Oliva, Olivella*
Family Harpidae: *Harpa*
Family Mitridae: *Mitra*
Superfamily Conacea (Toxoglossa)
Family Turridae
Family Conidae: *Conus, Lora*
Family Terebridae: *Terebra*

Subclass **Opisthobranchia**

Order Bullomorpha
Suborder Bullacea
Family Bullidae
Suborder Atyacea
Family Atyidae: *Haminoea*
Suborder Philinacea
Family Aglajidae: *Melanochlamys, Navanax*
Family Philinidae: *Philine*

Order Pyramidellomorpha
Family Pyramidellidae: *Odostomia*

Order Thecosomata
Suborder Euthecosomata
Family Spiratellidae: *Spiratella*
Family Cavoliniidae: *Cavrolinia, Creseis*
Suborder Pseudothecosomata
Family Cymbuliidae: *Gleba*

Order Gymnosomata
Family Clionidae: *Clione*

Order Aplysiomorpha
Family Aplysiidae: *Aplysia*

Order Pleurobranchomorpha
Suborder Pleurobranchacea
Family Pleurobranchidae: *Pleurobranchaea*

Order Sacoglossa
Suborder Elysiacea
Family Elysiidae: *Elysia*

Family Placobranchidae: *Placobranchus*
Family Stiligeridae: *Stiliger*
Family Limapontiidae: *Limapontia*
Family Oleidae: *Olea*

Order Nudibranchia
Suborder Dendronotacea
Family Tritonidae: *Tritonia*
Family Dendronotidae: *Dendronotus*
Suborder Anadoridoidea
Family Onchidoridae: *Onchidoris*
Suborder Eudorididae
Family Chromodorididae: *Casella*, *Chromodoris*
Family Archidoridae: *Archidoris*
Suborder Aeolidacea
Superfamily Euaeolidoidea
Family Pteraeolidiidae: *Pteraeolidia*
Family Aeolidiidae: *Aeolidia*

Subclass **Pulmonata**

Order Systellommatophora
Family Onchidiidae: *Onchidium*

Order Basommatophora
Family Siphonariidae: *Siphonaria*
Family Trimusculidae: *Gadinalea*
**Family* Ancylidae: *Ancylus*
**Family* Planorbidae: *Planorbis*

---

* *Freshwater.*

References

Andrews, E.B. (1976). 'The Fine Structure of the Heart of Some Prosobranch and Pulmonate Gastropods in Relation to Filtration', *Journal of Molluscan Studies, 42*, 199-216

Ansell, A.D. (1981). 'Experimental Studies of a Benthic Predator-Prey Relationship. I. Feeding, Growth, and Egg-collar Production in Long-term Cultures of the Gastropod Drill *Polinices alderi* (Forbes) Feeding on the Bivalve *Tellina tenuis* (da Costa)', *Journal of Experimental Marine Biology and Ecology, 56*, 235-55

Ansell, A.D. (1982). 'Experimental Studies of a Benthic Predator-Prey Relationship. II. Energetics of Growth and Reproduction, and Food-conversion Efficiencies, in Long-term Cultures of the Gastropod Drill *Polinices alderi* (Forbes) Feeding on the Bivalve *Tellina tenuis* da Costa', *Journal of Experimental Marine Biology and Ecology, 61*, 1-29

Ansell, A.D. and Trevallion, A. (1970). 'Brood Protection in the Stenoglossan Gastropod *Bullia mellanoides* (Deshayes)', *Journal of Natural History, 4*, 369-74

Bannister, J.V. (1974). 'The Respiration in Air and in Water of the Limpets *Patella caerulea* (L.) and *Patella lusitanica* (Gmelin)', *Comparative Biochemistry and Physiology, 49A*, 407-11

Bayne, B.L. and Scullard, C. (1978a). 'Rates of Oxygen Consumption by *Thais (Nucella) lapillus* (L.)', *Journal of Experimental Marine Biology and Ecology, 32*, 97-111

Bayne, B.L. and Scullard, C., (1978b). 'Rates of Feeding by *Thais (Nucella) lapillus* (L.)', *Journal of Experimental Marine Biology and Ecology, 32*, 113-29

Begon, M. and Mortimer, M. (1981). *Population Ecology: a Unified Study of Animals and Plants*, Blackwell Scientific Publications, Oxford

Behrens-Yamada, S. (1977). 'Geographic Range Limitation of the Intertidal Gastropods *Littorina sitkana* and *L. planaxis*', *Marine Biology, 39*, 61-5

Berger, E.M. (1977). 'Gene-Enzyme Variation in Three Sympatric Species of *Littorina*. II. The Roscoff Population with a Note on the Origin of North American *L. littorea*', *Biological Bulletin, 153*, 255-64

Bertalanffy, L. von (1957). 'Quantitative Laws in Metabolism and Growth', *Quarterly Review of Biology, 32*, 217-31

Bertness, M.D. (1984). 'Habitat and Community Modification by an Introduced Herbivorous Snail', *Ecology, 65*, 370-81

Bertness, M.D., Yund, P.O. and Brown, A.F. (1983). 'Snail Grazing and the Abundance of Algal Crusts on a Sheltered New England Rocky Beach', *Journal of Experimental Marine Biology and Ecology, 71*, 147-64

Black, R. (1978). 'Tactics of Whelks Preying on Limpets', *Marine Biology, 46*, 157-62

Blackmore, D.T. (1969). 'Studies of *Patella vulgata* L. II. Seasonal Variation in Biochemical Composition', *Journal of Experimental Marine Biology and Ecology, 3*, 231-45

Blundon, J.A. and Vermeij, G.J. (1983). 'Effect of Shell Repair on Shell Strength in the Gastropod *Littorina irrorata*', *Marine Biology, 76*, 41-5

Boddington, M.J. (1978). 'An Absolute Metabolic Scope for Activity', *Journal of Theoretical Biology, 75*, 443-9

Boggs, C.H., Rice, J.A., Kitchell, J.A. and Kitchell, J.F. (1984). 'Predation at a Snail's Pace: What's Time to a Gastropod?', *Oecologia (Berlin), 62*, 13-17

Boghen, A. and Farley, J. (1974). 'Phasic Activity in the Digestive Gland Cells of the Intertidal Prosobranch, *Littorina saxatilis* (Olivi) and its Relation to the Tidal Cycle', *Proceedings of the Malacological Society of London, 41*, 41-56

Bonaventura, C. and Bonaventura, J. (1983). 'Respiratory Pigments: Structure and

Function', in Hochachka, P.W. and Wilbur, K.M. (eds.), *The Mollusca, Vol. 2, Environmental Biology and Physiology*, Academic Press, London, pp. 1-50

Boss, J.K. (1971). 'Critical Estimate of the Number of Recent Mollusca', *Occasional Papers on Mollusks*, Museum of Comparative Zoology, Harvard, *3*, 81-135

Boyle, P.R., Sillar, M. and Bryceson, K. (1979). 'Water Balance and the Mantle Cavity Fluid of *Nucella lapillus* (L.) (Mollusca: Prosobranchia)', *Journal of Experimental Marine Biology and Ecology, 40*, 41-51

Branch, G.M. (1975). 'Mechanisms Reducing Intraspecific Competition in *Patella.* spp.: Migration, Differentiation and Territorial Behaviour', *Journal of Animal Ecology, 44*, 575-600

Branch, G.M. (1981). 'The Biology of Limpets: Physical Factors, Energy Flow, and Ecological Interactions', *Oceanography and Marine Biology Annual Review, 19*, 235-380

Branch, G.M. and Newell, R.C. (1978). 'A Comparative Study of Metabolic Energy Expenditure in the Limpets *Patella cochlear, P. oculus* and *P. granatina*', *Marine Biology, 49*, 351-61

Brenchley, G.A. (1982). 'Predation on Encapsulated Larvae by Adults: Effects of Introduced Species on the Gastropod *Ilyanassa obsoleta*', *Marine Ecology Progress Series, 9*, 255-62

Brenchley, G.A. and Carlton, J.T. (1983). 'Competitive Displacement of Native Mud Snails by Introduced Periwinkles in the New England Intertidal Zone', *Biological Bulletin, 165*, 543-58

Bretz, D.D. and Dimock, R.V. (1983). 'Behaviourally Important Characteristics of the Mucous Trail of the Marine Gastropod *Ilyanassa obsoleta* (Say)', *Journal of Experimental Marine Biology and Ecology, 71*, 181-91

Briggs, J.C. (1974), *Marine Zoogeography*, McGraw-Hill, New York

Broom, M.J. (1983). 'A Preliminary Investigation into Prey Species Preference by the Tropical Gastropods *Natica maculosa* Lamarck and *Thais carinifera* (Lamarck)', *Journal of Molluscan Studies, 49*, 43-52

Brown, A.C. (1982). 'The Biology of Sandy-beach Whelks of the Genus *Bullia* (Nassariidae)', *Oceanography and Marine Biology Annual Review, 20*, 309-61

Brown, A.C. and Da Silva, F.M. (1979). 'The Effects of Temperature on Oxygen Consumption in *Bullia digitalis* Meuschen (Gastropoda, Nassaridae)', *Comparative Biochemistry and Physiology, 62A*, 573-6

Brown, A.C. and Da Silva, F.M. (1984). 'Effects of Temperature on Oxygen Consumption in Two Closely-related Whelks from Different Temperature Regimes', *Journal of Experimental Marine Biology and Ecology, 84*, 145-53

Brown, A.C. and Trueman, J.M.S. (1982). 'Muscles that Push Snails Out of their Shells', *Journal of Molluscan Studies, 48*, 97-8

Brown, A.C., Ansell, A.D. and Trevallion, A. (1978). 'Oxygen Consumption by *Bullia (Dorsanum) melanoides* (Deshayes) and *Bullia digitalis* Meuschen (Gastropoda, Nassaridae) — an Example of Non-acclimation', *Comparative Biochemistry and Physiology, 61A*, 123-5

Bulloch, A.G.M. and Dorsett, D.A. (1979a). 'The Functional Morphology and Motor Innervation of the Buccal Mass of *Tritonia hombergi*', *Journal of Experimental Biology, 79*, 7-22

Bulloch, A.G.M. and Dorsett, D.A. (1979b). 'The Integration of the Patterned Output of Buccal Motoneurons during Feeding in *Tritonia hombergi*', *Journal of Experimental Biology, 79*, 23-40

Burgh, M.E. de and Singla, C.L. (1984). 'Bacterial Colonization and Endocytosis on the Gill of a New Limpet Species from a Hydrothermal Vent', *Marine Biology, 84*, 1-6

Calow, P. (1974). 'Evidence for Bacterial Feeding in *Planorbis contortus* Linn.', *Proceedings of the Malacological Society of London, 41*, 145-56

Calow, P. (1975). 'Defecation Strategies of Two Freshwater Gastropods, *Ancylus fluviatilis* (Mull) and *Planorbis contortus* Linn. (Pulmonata) with a Comparison of Field and Laboratory Estimates of Food Absorption Rate', *Oecologia (Berlin), 20,* 51-63
Calow, P. (1979). 'Why Some Metazoan Mucus Secretions Are More Susceptible to Microbial Attack than Others', *American Naturalist, 114,* p. 149
Calow, P. (1981). *Invertebrate Biology: a Functional Approach,* Croom Helm, London
Calow, P. (1983). 'Life-cycle Patterns and Evolution', in Russell-Hunter, W.D. (ed.), *The Mollusca, Vol. 6, Ecology,* Academic Press, New York, pp. 649-78
Carefoot, T.H. (1967a). 'Growth and Nutrition of *Aplysia punctata* Feeding on a Variety of Marine Algae', *Journal of the Marine Biological Association of the United Kingdom, 47,* pp. 565-89
Carefoot, T.H. (1967b). 'Growth and Nutrition of Three Species of Opisthobranch Molluscs', *Comparative Biochemistry and Physiology, 21,* pp. 627-52
Carriker, M.R. (1969). 'Excavation of Boreholes by the Gastropod *Urosalpinx*: an Analysis by Light and Scanning Electron Microscopy', *American Zoologist, 9,* pp. 917-33
Carriker, M.R. and Van Zandt, D. (1972). 'Predatory Behaviour of a Shell-boring Muricid Gastropod', in Winn, H.E. and Olla, B.L. (eds.), *Behaviour of Marine Animals: Current Perspectives in Research, Vol. 1, Invertebrates,* Plenum, New York, pp. 157-244
Cassidy, M.D. and Evans, S.M. (1981). 'Foraging Behaviour in Relation to Position on the Seashore in the Limpet *Patella vulgata*', *Animal Behaviour, 29,* pp. 300-1
Chaffee, C. and Strathmann, R.R. (1984). 'Constraints on Egg Masses. I. Retarded Development within Thick Egg Masses', *Journal of Experimental Marine Biology and Ecology, 84,* pp. 73-83
Chia, F.S. and Skeel, M. (1973). 'The Effect of Food Consumption on Growth, Fecundity, and Mortality in a Sacoglossan Opisthobranch, *Olea hansineensis*', *Veliger, 16,* pp.153-8
Christiansen, F.B. and Fenchel, T. (1977). 'Theories of Populations and Biological Communities', in *Ecological Studies: Analysis and Synthesis, 20,* Billings, W.D., Golley, F., Lange, O.L. and Olson J.J. (eds.), Springer-Verlag, Berlin, pp. 32-6
Clark, K.B. and Busacca, M. (1978). 'Feeding Specificity and Chloroplast Retention in Four Tropical Ascoglossa, with a Discussion of the Extent of Chloroplast Symbiosis and the Evolution of the Order', *Journal of Molluscan Studies, 44,* pp. 272-82
Clark, K.B. Busacca, M.B. and Stirts, H. (1979). 'Nutritional Aspects of Development of the Ascoglossan, *Elysia cauze*', in Stancyk, S.E. (ed.), *Reproductive Ecology of Marine Invertebrates, Belle W. Baruch Library in Marine Science, 9,* 11-24, University of South Carolina Press, Columbia
Clark, W.C. (1958). 'Escape Responses of Herbivorous Gastropods when Stimulated by Carnivorous Gastropods', *Nature (London), 181,* pp. 137-8
Connell, J.H. (1980). 'Diversity and the Coevolution of Competitors, or the Ghost of Competition Past', *Oikos, 35,* pp. 131-8
Connor, V.M. and Quinn, J.F. (1984). 'Stimulation of Food Species Growth by Limpet Mucus', *Science, 225,* pp. 843-4
Connover, R.J. and Lalli, C.M. (1972). 'Feeding and Growth in *Clione limacina* (Phillipps), a Pteropod Mollusc', *Journal of Experimental Marine Biology and Ecology, 9,* pp. 279-302
Cook, A., Bamford, O.S., Freeman, J.D.B. and Teideman, D.J. (1969). 'A Study of the Homing Habit of the Limpet', *Animal Behaviour, 17,* pp. 330-9
Creese, R.G. (1981). 'Patterns of Growth, Longevity and Recruitment of Intertidal

Limpets in New South Wales', *Journal of Experimental Marine Biology and Ecology, 51*, pp. 145-71
Crisp, D.J. (1974). 'Factors Influencing the Settlement of Marine Invertebrate Larvae', in Grant, P.T. and Mackie, A.M. (eds.), *Chemoreception in Marine Organisms*, Academic Press, New York, pp. 175-265
Crisp, D.J. (1976). 'The Role of the Pelagic Larva', in Spencer-Davies, P. (ed.), *Perspectives in Experimental Biology Vol. 1, Zoology*, Pergamon, Oxford, pp. 145-55
Crisp, M., Davenport, J. and Shumway, S.E. (1978). 'Effects of Feeding and of Chemical Stimulation on the Oxygen Uptake of *Nassarius reticulatus* (Gastropoda: Prosobranchia)', *Journal of the Marine Biological Association of the United Kingdom, 58*, pp. 387-90
Crisp, M., Gill, C.W. and Thompson, M.C. (1981). 'Ammonia Excretion by *Nassarius reticulatus* and *Buccinum undatum* (Gastropoda: Prosobranchia) during Starvation and after Feeding', *Journal of the Marine Biological Association of the United Kingdom, 61*, pp. 381-90
Currey, J.D. and Hughes, R.N. (1982). 'Strength of the Dogwhelk *Nucella lapillus* and the Winkle *Littorina littorea* from Different Habitats', *Journal of Animal Ecology, 51*, pp. 47-56
Currey, J.D. and Kohn, A.J. (1976). 'Fracture in the Crossed-lamellar Structure of *Conus* Shells', *Journal of Materials Science, 11*, pp. 1615-23
Currey, J.D. and Taylor, J.D. (1974). 'The Mechanical Behaviour of Some Molluscan Hard Tissues', *Journal of Zoology, London, 173*, pp. 395-406
Curtis, L.A. and Hurd, L.E. (1981). 'Crystalline Style Cycling in *Ilyanassa obsoleta* (Say) (Mollusca: Neogastropoda); Further Studies', *Veliger, 24*, pp. 91-102
Daguzan, J. and Razet, P. (1971). 'Les elements teminaux du catabolisme azote et leurs variations en function de la marée chez *Littorina littorea* (L.) adulte (Mollusque Mesogasteropode Littorinidae)', *Comptes Rendus hebdomadaires des Séances de l'Academie des Sciences, Paris, 272D*, pp. 2800-3
Davies, P.S. (1966). 'Physiological Ecology of *Patella*. I. The Effect of Body Size and Temperature on Metabolic Rate', *Journal of the Marine Biological Association of the United Kingdom, 46*, pp. 647-58
Davies, P.S. and Tribe, N.A. (1969). 'Temperature Dependence of Metabolic Rate in Animals', *Nature (London), 224*, pp. 723-4
Dawkins, R. (1976). *The Selfish Gene*, Oxford University Press, Oxford
Denny, M. (1980). 'Locomotion: the Cost of Gastropod Crawling', *Science, 208*, pp. 1208-12
Dogterom, G.E., Thijssen, R. and van Loenhout, H. (1985). 'Environmental and Hormonal Control of the Seasonal Egg-laying Period in Field Specimens of *Lymnaea stagnalis*', *General and Comparative Endocrinology, 57*, pp. 37-42
Doran, F.R. and McKenzie, D.S. (1972). 'Aerial and Aquatic Respiratory Responses to Temperature Variations in *Acmaea digitalis* and *Acmaea fenestrata*', *Veliger, 15*, pp. 38-52
Duerr, F.G. (1968). 'Excretion of Ammonia and Urea in Seven Species of Marine Prosobranch Snails', *Comparative Biochemistry and Physiology, 26*, pp. 1051-9
Dye, A.H. and McGwynne, L. (1980). 'The Effect of Temperature and Season on the Respiratory Rates of Three Psammolittoral Gastropods', *Comparative Biochemistry and Physiology, 66A*, pp. 107-11
Edwards, D.C. (1969). 'Predators on *Olivella biplicata*, Including a Species-specific Predator Avoidance Response', *Veliger, 11*, pp. 326-33
Edwards, D.C. and Huebner, J.D., (1977). 'Feeding and Growth Rates of *Polinices duplicatus* Preying on *Mya arenaria* at Barnstable Harbor, Massachusetts', *Ecology, 58*, pp. 1218-36
Ekaratne, S.U.K. and Crisp, D.J. (1982). 'Tidal Micro-growth Bands in Intertidal

Gastropod Shells, with an Evaluation of Band-dating Techniques', *Proceedings of the Royal Society, London B, 214*, pp. 305-23
Emlen, J.M. (1966). *Time, Energy and Risk in Two Species of Carnivorous Gastropods.* Ph.D. dissertation, University of Washington, Seattle
Fabens, A.J. (1965). 'Properties and Fitting of the von Bertalanffy Growth Curve', *Growth, 29*, pp. 265-89
Fawcett, M.H. (1984). 'Local and Latitudinal Variation in Predation on an Herbivorous Marine Snail', *Ecology, 65*, pp. 1214-30
Feare, C.J. (1971a). 'The Adaptive Significance of Aggregation Behaviour in the Dogwhelk *Nucella lapillus* (L.)', *Oecologia (Berlin), 7*, pp. 117-26
Feare, C.J. (1971b). 'Predation of Limpets and Dogwhelks by Oystercatchers', *Bird Study, 18*, pp. 121-9
Feder, H.M. (1972). 'Escape Responses in Marine Invertebrates', *Scientific American, 227*, pp. 92-100
Fenchel, T. (1975). 'Character Displacement and Coexistence in Mud Snails (Hydrobiidae)', *Oecologia (Berlin), 20*, pp. 19-32
Féral, C. (1979). 'Étude des facteurs regissant l'apparition d'un penis chez les femelles d'*Ocenebra erinacea* L. (Mollusque Gasteropode gonochorique) de la station d'Arcachon', *Comptes Rendus hebdomadaires des Séances de l'Academie des Sciences, Paris, 289*, pp. 331-4
Fields, J.H.A. (1983). 'Alternatives to Lactic Acid: Possible Advantages', *Journal of Experimental Zoology, 228*, pp. 445-57
Findley, A.M., Belisle, B.W. and Stickle, W.B. (1978). 'Effects of Salinity Fluctuations on the Respiration Rate of the Southern Oyster Drill *Thais haemastoma* and the Blue Crab *Callinectes sapidus*', *Marine Biology, 49*, pp. 59-67
Fioroni, P. (1966). 'Zur Morphologie und Embryogenese der Darmtraktes und der transitorischen Organe bei Prosobranchiern (Mollusca, Gastropoda), *Revue Suisse Zoologie, 73*, pp. 621-876
Fish, J.D. and Sharp, L. (1985). 'Population Dynamics, Growth and Reproduction of the Periwinkle, *Littorina neglecta* Bean', in Moore, P.G. and Seed, R. (eds), *The Ecology of Rocky Coasts*, Hodder and Stoughton, London, pp. 143-56
Fisher, R.A. (1930). *The Genetical Theory of Natural Selection*, Clarendon Press, Oxford
Fletcher, W.J. (1984). 'Variability in the Reproductive Effort of the Limpet *Cellana tramoserica*', *Oecologia (Berlin), 61*, pp. 259-64
Focardi, S., Deneubourg, J.L. and Chelazzi, G. (1985). 'How Shore Morphology and Orientation Mechanisms Can Affect the Spatial Organisation of Intertidal Molluscs', *Journal of Theoretical Biology, 112*, pp. 771-82
Frank, P.W. (1975). 'Latitudinal Variation in the Life-history Features of the Black Turban Snail *Tegula funebralis* (Prosobranchia: Trochidae)', *Marine Biology, 31*, pp. 181-92
Fretter, V. (1951). 'Some Observations on the British Cypraeids', *Proceedings of the Malacological Society, London, 29*, pp. 14-20
Fretter, V. (1975). '*Umbonium vestiarium*, a Filter-feeding Trochid', *Journal of Zoology, London, 177*, pp. 541-52
Fretter, V. and Graham, A. (1949). 'The Structure and Mode of Life of the Pyramidellidae, Parasitic Opisthobranchs', *Journal of the Marine Biological Association of the United Kingdom, 28*, pp. 493-532
Fretter, V. (1962). *British Prosobranch Molluscs*, Ray Society, London
Fukuyama, A. and Nybakken, J. (1984). 'Specialized Feeding in Mitrid Gastropods: Evidence from a Temperate Species, *Mitra idae* Melvill', *Veliger, 26*, pp. 96-100
Funke, W. (1968). 'Heimfindevermogen und Ortstreue bei *Patella* L. (Gastropoda, Prosobranchia)', *Oecologia (Berlin), 2*, pp. 19-142

Gäde, G., Carlsson, K-H. and Meinardus, G. (1984). 'Energy Metabolism in the Foot of the Marine Gastropod *Nassa mutabilis* during Environmental and Functional Anaerobiosis', *Marine Biology, 80*, pp. 49-56

Gainey, L.F. (1976). 'Locomotion in the Gastropoda: Functional Morphology of the Foot in *Neritina reclivata* and *Thais rustica*', *Malacologia, 15*, pp. 411-31

Gainey, L.F. and Wise, S.W. (1976). 'Functional Morphology of the Inner Surface of Gastropod (Mollusca) Shells', *Scanning Electron Microscopy/1976 (Part VIII). Proceedings of the Workshop on Zoological Applications of SEM, IIT Research Institute, Chicago*, April 1976

Ganaros, A.E. (1957). 'Marine Fungus Infecting Eggs and Embryos of *Urosalpinx cinerea*', *Science, 125*, p. 1194

Garrity, S.D. and Levings, S.C. (1981). 'A Predator-Prey Interaction between Two Physically and Biologically Constrained Tropical Rocky Shore Gastropods: Direct, Indirect and Community Effects', *Ecological Monographs, 51*, pp. 269-86

Garrity, S.D. and Levings, S.C. (1983). 'Homing to Scars as a Defence against Predators in the Pulmonate Limpet *Siphonaria gigas* (Gastropoda)', *Marine Biology, 172*, pp. 319-24

Garrity, S.D. and Levings, S.C. (1984). 'Aggregation in a Tropical Neritid', *Velinger, 27*, pp. 1-6

Geller, J.B. (1982). 'Chemically Mediated Avoidance Response of a Gastropod, *Tegula funebralis* (A. Adams), to a Predatory Crab, *Cancer antennarius* (Stimpson)', *Journal of Experimental Marine Biology and Ecology, 65*, pp. 19-27

Getting, P.A. (1983). 'Neuronal Control of Swimming in *Tritonia*', in Roberts, A. and Roberts, B.L. (eds), *Neural Origin of Rhythmic Movements. Symposium of the Society of Experimental Biology, 37*, pp. 89-128. Cambridge University Press, Cambridge

Ghiretti, F. (1966). 'Molluscan Hemocyanins', in Wilbur, K.M. and Yonge, C.M. (eds), *Physiology of Mollusca II*, Academic Press, New York, pp. 233-48

Gilmer, R.W. (1972). 'Free-floating Mucus Webs: a Novel Feeding Adaptation for the Open Ocean', *Science, 176*, pp. 1239-40

Gilmer, R.W. (1974). 'Some Aspects of Feeding in the Thecosomatous Pteropod Molluscs', *Journal of Experimental Marine Biology and Ecology, 15*, pp. 127-44

Gnaiger, E. (1983). 'Heat Dissipation and Energetic Efficiency in Animal Anoxibiosis: Economy Contra Power', *Journal of Experimental Zoology, 228*, pp. 471-90

Gohar, H.A. and Eisawy, A.M. (1967). 'The Egg Masses and Development of Five Rachiglossan Prosobranchs (from the Red Sea)', *Publications of the Marine Biological Station, Al-Ghardaqua, Egypt, 14*, pp. 216-66

Gonor, J.J. (1965). 'Predator-Prey Interactions between Two Marine Prosobranch Gastropods', *Veliger, 7*, pp. 228-32

Graham, A. (1938). 'On a Ciliary Process of Food-collecting in the Gastropod *Turritella communis* Risso', *Proceedings of the Zoological Society, London, 108*, pp. 453-63

Graham, A. (1982). 'A Note on *Cima minima* (Prosobranchia, Aclididae)', *Journal of Molluscan Studies, 48*, p. 232

Grahame, J. (1969). 'The Biology of *Berthelinia caribbea* Edmunds', *Bulletin of Marine Science of the Gulf and Caribbean, 19*, pp. 868-79

Grahame, J. (1973). 'Breeding Energetics of *Littorina littorea* (L.) (Gastropoda: Prosobranchiata)', *Journal of Animal Ecology, 42*, pp. 391-403

Grahame, J. (1977). 'Reproductive Effort and *r*- and *K*-selection in Two Species of *Lacuna* (Gastropoda: Prosobranchia)', *Marine Biology, 40*, pp. 217-24

Grenon, J.F. and Walker, G. (1981). 'The Tenacity of the Limpet, *Patella vulgata* L.: an Experimental Approach', *Journal of Experimental Marine Biology and*

Ecology, 54, pp. 277-308
Griffiths, C.L. (1981). 'Predation on the Bivalve *Choromytilus meridionalis* (Kr.) by the Gastropod *Natica (Tectonatica) tecta* Anton', *Journal of Molluscan Studies, 47*, pp. 112-20
Gunter, G. (1979). 'Studies on the Southern Oyster Borer, *Thais haemastoma*', *Gulf Research Report, 6*, pp. 249-60
Gurin, S. and Carr, W.E. (1971). 'Chemoreception in *Nassarius obsoletus*: the Role of Specific Stimulating Proteins', *Science (Washington, DC), 174*, pp. 293-5
Hadfield, M.G. (1963). 'The Biology of Nudibranch Larvae', *Oikos, 14*, pp. 85-95
Hadfield, M.G., Kay, E.A., Gillette, M.U. and Lloyd, M.C. (1972). 'The Vermetidae (Mollusca: Gastropoda) of the Hawaiian Islands', *Marine Biology, 12*, pp. 81-98
Hadlock, R.P. (1980). 'Alarm Response of the Intertidal Snail *Littorina littorea* (L.) to Predation by the Crab *Carcinus maenas* (L.)', *Biological Bulletin, 159*, pp. 269-79
Hall, J.G. and Feng, S.Y. (1976). 'Genital Variation among Connecticut Populations of the Oyster Drill *Urosalpinx cinerea* Say. (Prosobranchia; Muricidae)', *Veliger, 18*, pp. 318-21
Hall, S.J., Todd, C.D. and Gordon, A.D. (1982). 'The Influence of Ingestive Conditioning on the Prey Species Selection in *Aeolidia papillosa* (Mollusca: Nudibranchia)', *Journal of Animal Ecology, 51*, pp. 907-21
Hamner, W.M., Madin, L.P., Alldredge, A.L., Gilmer, R.W. and Hamner, P.P. (1975). 'Underwater Observations of Gelatinous Zooplankton: Sampling Problems, Feeding Biology, and Behavior', *Limnology and Oceanography, 20*, pp. 907-17
Hannaford-Ellis, C.J. (1983). 'Patterns of Reproduction in Four *Littorina* Species', *Journal of Molluscan Studies, 49*, pp. 98-106
Hargens, A.R. and Shabica, S.V. (1973). 'Protection against Lethal Freezing Temperatures by Mucus in an Antarctic Limpet', *Cryobiology, 10*, pp. 331-7
Hawkins, S.J. (1981). 'The Influence of *Patella* Grazing on the Fucoid/Barnacle Mosaic on Moderately Exposed Rocky Shores', *Kieler Meeresforschungen, 5*, pp. 537-43
Heath, D.J. (1977). 'Simultaneous Hermaphroditism; Cost and Benefit', *Journal of Theoretical Biology, 64*, pp. 363-73
Heath, D.J. (1985). 'Whorl Overlap and the Economical Construction of the Gastropod Shell', *Biological Journal of the Linnaean Society, 24*, pp. 289-307
Hemmingsen, A.M. (1960). 'Energy Metabolism as Related to Body Size and Respiratory Surfaces, and its Evolution', *Reports of the Steno Memorial Hospital and the Nordisk Insulinlaboratorium, 9*, pp. 1-110
Hickman, C.S. (1984). 'Implications of Radular Tooth-row Functional Integration for Archaeogastropod Systematics', *Malacologia, 25*, pp. 143-60
Hickman, C.S. and Morris, T.E. (1985). 'Gastropod Feeding Tracks as a Source of Data in Analysis of the Functional Morphology of Radulae', *Veliger, 27*, pp. 357-65
Himmelman, J.H. (1978). 'Phytoplankton as a Stimulus for Spawning in Three Marine Invertebrates', *Journal of Experimental Marine Biology and Ecology, 20*, pp. 199-214
Hoagland, K.E. (1978). 'Protandry and the Evolution of Environmentally-mediated Sex Change: a Study of the Mollusca', *Malacologia, 17*, pp. 365-91
Horn, H.S. (1978). 'Optimal Tactics of Reproduction and Life-history', in Krebs, J.R. and Davies, N.B. (eds), *Behavioural Ecology: an Evolutionary Approach*, Blackwell Scientific Publications, Oxford, pp. 411-29
Houbrick, R.S. (1973). 'Studies on the Reproductive Biology of the Genus

Cerithium (Gastropoda: Prosobranchia) in the Western Atlantic', *Bulletin of Marine Science, 23,* pp. 875-904

Houbrick, R.S. and Fretter, V. (1969). 'Some Aspects of the Functional Anatomy and Biology of *Cymatium* and *Bursa*', *Proceedings of the Malacological Society of London, 38,* pp. 423-81

Houlihan, D.F. (1979). 'Respiration in Air and Water of Three Mangrove Snails', *Journal of Experimental Marine Biology and Ecology, 41,* pp. 143-61

Houlihan, D.F., Innes, A.J. and Day, D.G. (1981). 'The Influence of Mantle Cavity Fluid on the Aerial Oxygen Consumption of Some Intertidal Gastropods', *Journal of Experimental Marine Biology and Ecology, 49,* pp. 57-68

Huebner, J.D. (1973). 'The Effect of Body Size and Temperature on the Respiration of *Polinices duplicatus*', *Comparative Biochemistry and Physiology, 44A,* pp. 1185-97

Hughes, H.P.I. (1985). 'Radula Teeth of Hong Kong Caecidae', *Proceedings of the 2nd International Workshop of Malacology,* 1983, Hong Kong University Press, Hong Kong (in press)

Hughes, R.N. (1971a). 'Ecological Energetics of *Nerita* (Archaeogastropoda, Neritacea) Populations on Barbados, West Indies', *Marine Biology, 11,* pp. 12-22

Hughes, R.N. (1971b). 'Ecological Energetics of the Keyhole Limpet *Fissurella barbadensis* Gmelin', *Journal of Experimental Marine Biology and Ecology, 6,* pp. 167-78

Hughes, R.N. (1972). 'Annual Production of Two Nova Scotian Populations of *Nucella lapillus* (L.)', *Oecologia (Berlin), 8,* pp. 356-70

Hughes, R.N. (1978). 'The Biology of *Dendropoma corallinaceum* and *Serpulorbis natalensis,* Two South African Vermetid Gastropods', *Zoological Journal of the Linnean Society, 64,* pp. 111-27

Hughes, R.N. (1979a). 'Coloniality in Vermetidae (Gastropoda)', in Larwood, G. and Rosen, B.R. (eds), *Biology and Systematics of Colonial Organisms. Systematics Association Special Volume No. 11,* pp. 243-53

Hughes, R.N. (1979b). 'South African Populations of *Littorina rudis*', *Zoological Journal of the Linnean Society, 65,* pp. 119-126

Hughes, R.N. (1980). 'Optimal Foraging Theory in the Marine Context', *Oceanography and Marine Biology Annual Review, 18,* pp. 423-81

Hughes, R.N. (1985a). 'Laboratory Observations on the Feeding Behaviour, Reproduction and Morphology of *Galeodea echinophora* (Gastropoda: Cassidae)', *Zoological Journal of the Linnean Society,* in press

Hughes, R.N. (1985b). 'Feeding Behaviour of the Sessile Gastropod *Tripsycha tulipa* (Vermetidae)', *Journal of Molluscan Studies,* in press

Hughes, R.N. (1985c). 'Predatory Behaviour of *Natica unifasciata* Feeding Intertidally on Gastropods', *Journal of Molluscan Studies,* in press

Hughes, R.N. and Answer, P. (1982). 'Growth, Spawning and Trematode Infection of *Littorina littorea* (L.) from an Exposed Shore in North Wales', *Journal of Molluscan Studies, 48,* pp. 321-30

Hughes, R.N. and Dunkin, S.de B. (1984a). 'Behavioural Components of Prey Selection by Dogwhelks, *Nucella lapillus* (L.), Feeding on Mussels, *Mytilus edulis* L., in the Laboratory', *Journal of Experimental Marine Biology and Ecology, 77,* pp. 45-68

Hughes, R.N. and Dunkin, S.de B. (1984b). 'Effect of Dietary History on Selection of Prey, and Foraging Behaviour among Patches of Prey, by the Dogwhelk, *Nucella lapillus* (L.)', *Journal of Experimental Marine Biology and Ecology, 79,* pp. 159-72

Hughes, R.N. and Elner, R.W. (1979). 'Tactics of a Predator, *Carcinus maenas,* and Morphological Responses of the Prey, *Nucella lapillus*', *Journal of Animal*

Ecology, 48, pp. 65-78
Hughes, R.N. and Hughes, H.P.I. (1981). 'Morphological and Behavioural Aspects of Feeding in the Cassidae (Tonnacea, Mesogastropoda)', *Malacologia, 20*, pp. 385-402
Hughes, R.N. and Roberts, D.J. (1980a). 'Reproductive Effort of Winkles (*Littorina* spp.) with Contrasted Methods of Reproduction', *Oecologia (Berlin), 47*, pp. 130-6
Hughes, R.N. and Roberts, D.J. (1980b). 'Growth and Reproductive Rates of *Littorina neritoides* (L.) in North Wales', *Journal of the Marine Biological Association of the United Kingdom, 60*, pp. 591-9
Hughes, R.N. and Roberts, D.J. (1981). 'Comparative Demography of *Littorina rudis, L. nigrolineata* and *L. neritoides* on Three Contrasted Shores in North Wales', *Journal of Animal Ecology, 50*, pp. 251-68
Hutchinson, G.E. (1959). 'Homage to Santa Rosalia, or Why Are There So Many Kinds of Animals?', *American Naturalist, 93*, pp. 145-59
Hylleberg, J. (1976). 'Resource Partitioning on Basis of Hydrolytic Enzymes in Deposit-feeding Mud Snails (Hydrobiidae). II. Studies on Niche Overlap', *Oecologia (Berlin), 23*, pp. 115-25
Innes, A.J. (1984). 'The Effects of Aerial Exposure and Desiccation on the Oxygen Consumption of the Intertidal Limpets', *Veliger, 27*, pp. 134-9
Jablonski, D. and Lutz, R.A. (1980). 'Molluscan Larval Shell Morphology: Ecological and Paleontological Applications', in Rhoads, D.C. and Lutz, R.A. (eds), *Skeletal Growth of Aquatic Organisms*, Plenum, New York, pp. 323-77
James, B.L. (1968). 'The Distribution and Keys of Species in the Family Littorinidae and their Digenean Parasites, in the Region of Dale, Pembrokeshire', *Field Studies, 2*, pp. 615-50
Janson, K. and Ward, R.D. (1984). 'Microgeographic Variation in Allozyme and Shell Characters in *Littorina saxatilis* Olivi (Prosobranchia: Littorinidae)', *Biological Journal of the Linnaean Society, 32*, pp. 289-307
Jensen, K. (1975). 'Food Preference and Food Consumption in Relation to Growth in *Limapontia capitata* (Opisthobranchia, Sacoglossa)', *Ophelia, 14*, pp. 1-14
Jones, N.S. (1948). 'Observations and Experiments on the Biology of *Patella vulgata* at Port St. Mary, Isle of Man', *Proceedings and Transactions of the Liverpool Biological Society, 56*, pp. 60-77
Joosse, J. (1979). 'Evolutionary Aspects of the Endocrine System and of the Hormonal Control of Reproduction of Molluscs', in Barrington, E.J.W. (ed.), *Hormones and Evolution, 1*, Academic Press, New York, pp. 119-57
Jørgensen, C.B. (1952). 'On the Relation between Water Transport and Food Requirements in Some Marine Filter-feeding Invertebrates', *Biological Bulletin, 103*, pp. 356-63
Jørgensen, C.B., Kiørbae, T., Møhlenberg, F. and Riisgaard, H.U. (1984). 'Ciliary and Mucus-net Filter Feeding, with Special Reference to Fluid Mechanical Characteristics', *Marine Ecology Progress Series, 15*, pp. 283-92
Kandel, E.R. (1976). *Cellular Basis of Behavior. An Introduction to Behavioral Neurophysiology*, W.H. Freeman, San Francisco
Kemp, P. and Bertness, M.D. (1984). 'Snail Shape and Growth Rates: Evidence for Plastic Shell Allometry in *Littorina littorea*', *Proceedings of the National Academy of Science, 81*, pp. 811-13
Kempf, S.C. (1980). 'Symbiosis between Zooxanthellae and Three species of Nudibranchs', *American Zoologist, 20*, p. 777
Kempf, S. (1981). 'Long-lived Larvae of the Gastropod *Aplysia juliana*: Do they Disperse and Metamorphose or Just Slowly Fade Away?', *Marine Ecology Progress Series, 6*, pp. 61-5
Kerstitch, A. (1984). 'Killer Cone', *Opisthobranch, 16*, p. 13

Kingston, R.S. (1968). 'Anatomical and Oxygen Electrode Studies of Respiratory Surfaces and Respiration in *Acmaea*', in Abbott, D.P., Epel, D., Phillips, J.H., Abbott, I.A. and Stohler, R. (eds), *The Biology of Acmaea, Veliger, 11,* (Supplement), pp. 73-8

Kitchell, J.A., Boggs, C.H., Kitchell, J.E. and Rice, J.A. (1981). 'Prey Selection by Naticid Gastropods: Experimental Tests and Application to the Fossil Record', *Paleobiology, 7,* pp. 533-52

Kitching, R.L. and Pearson, J. (1981). 'Prey Location by Sound in a Predatory Intertidal Gastropod', *Marine Biology Letters, 2,* pp. 313-33

Kitting, C.L. (1979). 'The Use of Feeding Noises to Determine the Algal Foods Being Consumed by Individual Intertidal Molluscs', *Oecologia (Berlin), 40,* pp. 1-17

Kofoed, L.H. (1975a). 'The Feeding Biology of *Hydrobia ventrosa* (Montagu). I. The Assimilation of Different Components of the Food', *Journal of Experimental Marine Biology and Ecology, 19,* pp. 233-41

Kofoed, L.H. (1975b). 'The Feeding Biology of *Hydrobia ventrosa* (Montagu). 2. Allocation of the Components of the Carbon-budget and the Significance of the Secretion of Dissolved Organic Material', *Journal of Experimental Marine Biology and Ecology, 19,* pp. 243-56

Kohn, A.J. (1961). 'Chemoreception in Gastropod Molluscs', *American Zoologist, 1,* pp. 291-308

Kohn, A.J. (1967). 'Environmental Complexity and Species Diversity in the Gastropod Genus *Conus* on Indo-West Pacific Reef Platforms', *American Naturalist, 101,* pp. 251-9

Kohn, A.J. (1968). 'Microhabitats, Abundance and Food of *Conus* on Atoll Reefs in the Maldive and Chagos Islands', *Ecology, 49,* pp. 1046-62

Kohn, A.J. (1978). 'Ecological Shift and Release in an Isolated Population: *Conus miliaris* at Easter Island', *Ecological Monographs, 48,* pp. 323-36

Kohn, A.J. (1983). 'Feeding Biology of Gastropods', in Saleuddin, A.S.M. and Wilbur, K.M. (eds), *The Mollusca, Vol. 5, Physiology, Part 2,* Academic Press, New York, pp. 1-63

Kohn, A.J. and Waters, V.L. (1966). 'Escape Responses of Three Herbivorous Gastropods to the Predatory Gastropod *Conus textile*', *Animal Behaviour, 14,* pp. 340-5

Kohn, A.J., Myers, E.R. and Meenakshi, V.R. (1979). 'Interior Remodeling of the Shell by a Gastropod Mollusc', *Proceedings of the National Academy of Science, USA, 76,* pp. 3406-10

Kupferman, I. (1974). 'Dissociation of the Appetitive and Consumatory Phases of Feeding Behavior in *Aplysia*: a Lesion Study', *Behavioral Biology, 10,* pp. 89-97

Laird, A.K., Tyler, S.A. and Barton, A.D. (1965). 'Dynamics of Normal Growth', *Growth, 29,* pp. 233-48

Lalli, C.M. (1970). 'Structure and Function of the Buccal Apparatus of *Clione limacina* (Phillipps) with a Review of Feeding in Gymnosomatous Pteropods', *Journal of Experimental Marine Biology and Ecology, 4,* pp. 101-18

Lassen, H.H. (1979). 'Reproductive Effort in Danish Mudsnails (Hydrobiidae)', *Oecologia (Berlin), 40,* pp. 365-9

Le Gall, S. and Streiff, W. (1975). 'Protandric Hermaphroditism in Prosobranch Gastropods', in Reinboth, R. (ed.), *Intersexuality in the Animal Kingdom,* Springer-Verlag, Berlin, pp. 170-8

Lever, J. and Boer, H.H. (1983) (eds). *Molluscan Neuroendocrinology,* Proceedings of the International Minisymposium on Molluscan Endocrinology, held in the Department of Biology, Free University, Amsterdam, Netherlands, 16-20 August 1982, North-Holland Publishing Company, Amsterdam

Levington, J.S. (1982). 'The Body Size-Prey Size Hypothesis: the Adequacy of

Body Size as a Vehicle for Character Displacement', *Ecology, 63*, pp. 869-72
Levington, J.S. and Bianchi, T.S. (1981a). 'Nutrition and Food Limitation of Deposit-feeders. I. The Role of Microbes in the Growth of Mud Snails (Hydrobiidae)', *Journal of Marine Research, 39*, pp. 531-45
Levington, J.S. and Bianchi, T.S. (1981b). 'Nutrition and Food Limitation of Deposit-feeders. II. Differential Effects of *Hydrobia totteni* and *Ilyanassa obsoleta* on the Microbial Community', *Journal of Marine Research, 39*, pp. 547-56
Levington, P.J. (1978). 'Resource Partitioning by Predatory Gastropods of the Genus *Conus* on Subtidal Indo-Pacific Coral Reefs: the Significance of Prey Size', *Ecology, 59*, pp. 614-31
Lewis, J.B. (1971). 'Comparative Respiration of Some Tropical Intertidal Gastropods', *Journal of Experimental Marine Biology and Ecology, 6*, pp. 101-8
Lindberg, D.R. and Dwyer, K.R. (1983). 'The Topography, Formation and Role of the Home Depression of *Collisella scabra* (Gould)', *Veliger, 25*, pp. 229-34
Linsley, R.M. and Javidpour, M. (1980). 'Episodic Growth in Gastropods', *Malacologia, 20*, pp. 153-60
Lopez, G.R. and Cheng, I.J. (1983). 'Synoptic Measurements of Ingestion Rate, Ingestion Selectivity, and Absorption Efficiency of Natural Foods in the Deposit-feeding Molluscs *Nucula annulata* (Bivalvia) and *Hydrobia totteni* (Gastropoda)', *Marine Ecology Progress Series, 11*, pp. 55-62
Lubchenco, J. (1978). Plant Species Diversity in a Marine Intertidal Community: Importance of Herbivore Food Preference and Algal Competitive Abilities', *American Naturalist, 112*, pp. 23-9
Lützen, J. (1972). 'Studies on Parasitic Gastropods from Echinoderms. II. On *Stilifer* Broderip, with Special Reference to the Structure of the Sexual Apparatus and the Reproduction', *Det Kongelige Danske Videnskaberies Selskab Biologiske Skrifter, 19*, pp. 1-18
Lützen, J. (1979). 'Studies on the Life History of *Enteroxenos bonnevie*, a Gastropod Endoparasitic in Aspidochirote Holothurians', *Ophelia, 18*, pp. 1-51
Macé, A.M. and Ansell, A.D. (1982). 'Respiration and Nitrogen Excretion of *Polinices catena* (da Costa) (Gastropoda: Naticidae)', *Journal of Experimental Marine Biology and Ecology, 60*, pp. 275-92
MacArthur, R.H. and Wilson, E.O. (1967). *The Theory of Island Biogeography*, Princeton University Press, Princeton
McFarlane, I.D. (1981). 'In the Intertidal Homing Gastropod *Onchidium verruculatum* (Cuv.) the Outward and Homeward Trails Have a Different Information Content', *Journal of Experimental Marine Biology and Ecology, 51*, pp. 207-18
McKillup, S.C. and Butler, A.J. (1979). 'Modification of Egg Production and Packaging in Response to Food Availability by *Nassarius pauperatus*', *Oecologia (Berlin), 43*, pp. 221-31
McMahon, R.F. and Russell-Hunter, W.D. (1977). 'Temperature Relations of Aerial and Aquatic Respiration in Six Littoral Snails in Relation to their Vertical Zonation', *Biological Bulletin, 152*, pp. 182-98
Mapstone, B.D., Underwood, A.J. and Creese, R.G. (1984). 'Experimental Analysis of the Commensal Relation between Intertidal Gastropods *Patelloida mufria* and the Trochid *Austrocochlea constricta*', *Marine Ecology Progress Series, 17*, pp. 85-100
Margolin, A.S. (1964). 'A Running Response of *Acmaea* to Seastars', *Ecology, 45*, pp. 191-3
Markel, R.P. (1976). 'Some Biochemical Responses to Temperature Acclimation in the Limpet, *Acmaea limatula* Carpenter (1864)', *Comparative Biochemistry*

and Physiology, 53B, pp. 81-4
Mason, A.Z. and Nott, J.A. (1980). 'The Association of the Blood Vessels and the Excretory Epithelium in the Kidney of *Littorina littorea* (L.) (Mollusca, Gastropoda)', *Marine Biology Letters, 1*, pp. 355-65
Meglitsch, P.A. (1972). *Invertebrate Zoology*, Oxford University Press, London
Menge, B.L. (1978). 'Predator Activity in a Rocky Intertidal Community. Effect of an Algal Covering, Wave Action and Desiccation on Predator Feeding Rate', *Oecologia (Berlin), 34*, pp. 17-35
Menge, J. (1974). 'Prey Selection and Foraging Period of the Predaceous Rocky Intertidal Snail, *Acanthina punctulata*', *Oecologia (Berlin), 17*, pp. 293-316
Merdsoy, B. and Farley, J. (1973). 'Phasic Activity in the Digestive Gland Cells of the Marine Prosobranch Gastropod, *Littorina littorea* (L.)', *Proceedings of the Malacological Society of London, 40*, pp. 473-82
Miller, S.L. (1974). 'The Classification, Taxonomic Distribution, and Evolution of Locomotor Types among Prosobranch Gastropods', *Proceedings of the Malacological Society of London, 41*, pp. 233-73
Moran, M.J. (1985). 'Distribution and Dispersion of the Predatory Intertidal Gastropod, *Morula marginalba* (Muricidae)', *Marine Ecology Progress Series, 22*, pp. 41-52
Morton, J.E. (1951). 'The Structure and Adaptations of the New Zealand Vermetidae. parts I-III', *Transactions of the Royal Society of New Zealand, 79*, pp. 1-51
Morton, J.E. (1967). *Molluscs*, Hutchinson, London
Mpitsos, G.J. and Davis, W.J. (1973). 'Learning: Classical and Avoidance Conditioning in the Mollusc Pleurobranchaea', *Science (Washington, DC), 180*, pp. 317-20
Mpitsos, G.J., Collins, S.D. and McClellan, A.D. (1978). 'Learning: a Model System for Physiological Studies', *Science (Washington, DC), 191*, pp. 497-506
Murdoch, R.C. and Shumway, S.E. (1980). 'Oxygen Consumption of Six Species of Chitons in Relation to their Position on the Shore', *Ophelia, 19*, pp. 127-44
Murdoch, W.W. (1969). 'Switching in General Predators: Experiments on Predator Specificity and Stability of Prey Populations', *Ecological Monographs, 39*, pp. 335-54
Muscatine, L. and Greene, R.W. (1973). 'Chloroplasts and Algae as Symbionts in Molluscs', *International Review of Cytology, 36*, pp. 137-69
Naylor, E. (1976). 'Rhythmic Behaviour and Reproduction in Marine Animals', in Newell, R.C. (ed.), *Adaptations to Environment: Essays on the Physiology of Marine Animals*, Butterworths, London, pp. 393-429
Newell, R.C. (1965). 'The Role of Detritus in the Nutrition of Two Marine Deposit-feeders, the Prosobranch *Hydrobia ulvae* and the Bivalve *Macoma balthica*', *Proceedings of the Zoological Society of London, 144*, pp. 25-45
Newell, R.C. (1979). *Biology of Intertidal Animals*, Marine Ecological Surveys, Faversham, Kent
Newell, R.C. (1983). *Current Contents, 48*, p. 20
Newell, R.C. and Branch, G.M. (1980). 'The Influence of Temperature on the Maintenance of Metabolic Energy Balance in Marine Invertebrates', *Advances in Marine Biology, 17*, pp. 329-96
Newell, R.C. and Kofoed, L.H. (1977). 'The Energetics of Suspension Feeding in the Gastropod *Crepidula fornicata* L.', *Journal of the Marine Biological Association of the United Kingdom, 51*, pp. 161-80
Newell, R.C. and Northcroft, H.R. (1967). 'A Re-interpretation of the Effect of Temperature on the Metabolism of Certain Marine Invertebrates', *Journal of Zoology, 151*, pp. 277-98
Newell, R.C. and Pye, V.I. (1971). 'Quantitative Aspects of the Relationship between Metabolism and Temperature in the Winkle *Littorina littorea* (L.)',

Comparative Biochemistry and Physiology, 38B, pp. 635-50
Nicotri, M.E. (1977). 'Grazing Effects of Four Marine Intertidal Herbivores on the Microflora', *Ecology, 58*, pp. 1020-32
Nielsen, C. (1975). 'Observations on *Buccinum undatum* L. Attacking Bivalves and on Prey Responses, with a Short Review on Attack Methods of other Prosobranchs', *Ophelia, 13*, pp. 87-108
Ockelmann, K.W. and Nielsen, C. (1981). 'On the Biology of the Prosobranch *Lacuna parva* in the Oresund', *Ophelia, 20*, pp. 1-16
Odum, E.P. and Smalley, A.E. (1959). 'Comparison of Population Energy Flow of a Herbivorous and a Deposit-feeding Invertebrate in a Salt Marsh Ecosytem', *Proceedings of the National Academy of Science, USA, 45*, pp. 617-22
Ordzie, C.J. and Garofolo, G.C. (1980). 'Predation, Attack Success, and Attraction to the Bay Scallop, *Argopecten irradians* (Lamarck) by the Oyster Drill, *Urosalpinx cinerea* (Say)', *Journal of Experimental Marine Biology and Ecology, 47*, pp. 95-100
Paine, R.T. (1963). 'Feeding Rate of a Predaceous Gastropod, *Pleuroploca gigantea*', *Ecology, 44*, pp. 63-73
Paine, R.T. 'Natural History, Limiting Factors and Energetics of the Opisthobranch *Navanax inermis*', *Ecology, 46*, pp. 603-19
Paine, R.T. (1966). 'Function of Labial Spines, Composition of Diet, and Size of Certain Marine Gastropods', *Veliger, 9*, pp. 17-24
Paine, R.T. (1971). 'Energy Flow in a Natural Population of the Herbivorous Gastropod *Tegula funebralis*', *Limnology and Oceanography, 16*, pp. 86-98
Palmer, A.R. (1979). 'Fish Predation and the Evolution of Gastropod Shell Sculpture: Experimental and Geographic Evidence', *Evolution, 33*, pp. 697-713
Palmer, A.R. (1981). 'Do Carbonate Skeletons Limit the Rate of Body Growth?', *Nature, London, 292*, pp. 150-2
Palmer, A.R. (1982a). 'Growth in Marine Gastropods: a Non-destructive Technique for Independently Measuring Shell and Body Weight', *Malacologia, 23*, pp. 63-73
Palmer, A.R. (1982b). 'Predation and Parallel Evolution: Recurrent Parietal Plate Reduction in Balanomorph Barnacles', *Paleobiology, 8*, pp. 31-44
Palmer, A.R. (1983a). 'Growth Rate as a Measure of Food Value in Thaidid Gastropods: Assumptions and Implications for Prey Morphology and Distribution', *Journal of Experimental Marine Biology and Ecology, 73*, pp. 95-124
Palmer, A.R. (1983b). 'Relative Cost of Producing Skeletal Organic Matrix versus Calcification: Evidence from Marine Gastropods', *Marine Biology, 75*, pp. 287-92
Palmer, A.R. (1984). 'Prey Selection by Thaidid Gastropods: Some Observational and Experimental Field Tests of Foraging Models', *Oecologia (Berlin), 62*, pp. 162-72
Palmer, A.R. and Strathmann, R.R. (1981). 'Scale of Dispersal in Varying Environments and its Implications for Life Histories of Marine Invertebrates', *Oecologia (Berlin), 48*, pp. 308-18
Parry, G.D. (1978). 'Effects of Growth and Temperature Acclimation on Metabolic Rate in the Limpet, *Cellana tramoserica* (Gastropoda: Patellidae)', *Journal of Animal Ecology, 47*, pp. 351-68
Parry, G.D. (1984). 'The Effect of Food Deprivation on Seasonal Changes in the Metabolic Rate of the Limpet, *Cellana tramoserica*', *Comparative Biochemistry and Physiology, 77A*, pp. 663-8
Patterson, C.M. (1969). 'Chromosomes of Molluscs', *Proceedings of the Symposium on Mollusca, Marine Biological Association of India, Cochin, Part 2*, pp. 635-86

Pechenik, J.A. (1978a). 'Winter Reproduction in the Gastropod *Nassarius trivittatus*', *Veliger, 21,* pp. 297-8
Pechenik, J.A. (1978b). 'Adaptations to Intertidal Development: Studies on *Nassarius obsoletus*', *Biological Bulletin, 154,* pp. 282-91
Pechenik, J.A. (1979). 'Role of Encapsulation in Invertebrate Life Histories', *American Naturalist, 114,* pp. 859-70
Pechenik, J.A. (1980). 'Growth and Energy Balance during the Larval Lives of Three Prosobranch Gastropods', *Journal of Experimental Marine Biology and Ecology, 44,* pp. 1-28
Pechenik, J.A. (1983). 'Egg Capsules of *Nucella lapillus* (L.) Protect against Low Salinity Stress', *Journal of Experimental Marine Biology and Ecology, 71,* pp. 165-79
Pechenik, J.A. and Fisher, N.S., (1979). 'Feeding Assimilation, and Growth of Mud Snail Larvae, *Nassarius obsoletus* (Say), on Three Different Diets', *Journal of Experimental Marine Biology and Ecology, 38,* pp. 57-80
Pechenik, J.A., Chang, S.C. and Lord, A. (1984). 'Encapsulated Development of the Marine Prosobranch Gastropod *Nucella lapillus*', *Marine Biology, 78,* pp. 223-9
Perron, F.E. (1981). 'The Partitioning of Reproductive Energy between Ova and Protective Capsules in Marine Gastropods of the Genus *Conus*', *American Naturalist, 118,* pp. 110-18
Perron, F.E. and Corpuz, G.C. (1982). 'Costs of Parental Care in the Gastropod *Conus pennaceus*: Age-specific Changes and Physical Constraints', *Oecologia (Berlin), 55,* pp. 319-24
Peters, R.H. (1983). *The Ecological Implications of Body Size,* Cambridge University Press, Cambridge
Petpiroon, S. and Morgan, E. (1983). 'Observations on the Tidal Activity Rhythm of the Periwinkle *Littorina nigrolineata* (Gray)', *Marine Behaviour and Physiology, 9,* pp. 171-92
Phillips, D.W. (1975). 'Distance Chemoreception-triggered Avoidance Behavior of the Limpets *Acmaea (Collisella) limatula* and *Acmaea (Notoacmea) scutum* to the Predatory Starfish *Pisaster ochraceus*', *Journal of Experimental Zoology, 191,* pp. 199-210
Phillips, D.W. (1976). 'The Effect of a Species-specific Avoidance Response to Predatory Starfish on the Intertidal Distribution of Two Gastropods', *Oecologia (Berlin), 23,* pp. 83-94
Phillips, D.W. (1981). 'Life-history Features of the Marine Intertidal Limpet *Notoacmea scutum* (Gastropoda) in Central California', *Marine Biology, 64,* pp. 95-103
Pianka, E.R. (1970). 'On "*r*" and "*K*" Selection', *American Naturalist, 104,* pp. 592-7
Ponder, W.F. (1973). 'The Origin and Evolution of the Neogastropoda', *Malacologia, 12,* pp. 295-338
Pratt, D.M. (1976). 'Intraspecific Signalling of Hunting Success or Failure in *Urosalpinx cinerea*', *Journal of Experimental Marine Biology and Ecology, 21,* pp. 7-9
Prosser, C.L. (1973). *Comparative Animal Physiology,* 3rd Edn, W.B. Saunders, Philadelphia
Purchon, R.D. (1977). *The Biology of the Mollusca,* 2nd Edn, Pergamon Press, London
Pyke, G.H. (1978). 'Are Animals Efficient Harvesters?', *Animal Behaviour, 26,* pp. 241-50
Race, M.S. (1982). 'Competitive Displacement and Predation between Introduced and Native Mud Snails', *Oecologia (Berlin),* pp. 337-47
Raffaelli, D.G. (1978). 'The Relationship between Shell Injuries, Shell Thickness

and Habitat Characteristics of the Intertidal Snail *Littorina rudis* Maton', *Journal of Molluscan Studies, 44,* pp. 166-70
Raftery, R.E. (1983). '*Littorina* Trail Following: Sexual Preference, Loss of Polarized Information, and Trail Alterations', *Veliger, 25,* pp. 378-82
Ralph, R. and Maxwell, J.G.H. (1977). 'The Oxygen Consumption of the Antarctic Limpet *Nacella (Patinigera) concinna*', *Bulletin of the British Antarctic Survey, 45,* pp. 19-23
Rex, M.A. (1979). '*r*- and *K*-selection in a Deep-sea Gastropod', *Sarsia, 64,* pp. 29-32
Reynolds, W.W. and Reynolds, L.J. (1977). 'Zoogeography and the Predator-Prey 'Arms Race': a Comparison of *Eriphia* and *Nerita* Species from Three Faunal Regions', *Hydrobiologia, 56,* pp. 63-7
Riel, A. (1975). 'Effect of Trematodes on Survival of *Nassarius obsoletus* (Say)', *Proceedings of the Malacological Society, London, 41,* pp. 527-8
Robertson, A.I. (1979). 'The Relationship between Annual Production eg: Biomass Ratios and Lifespan for Marine Macrobenthos', *Oecologia (Berlin), 38,* pp. 193-202
Robertson, R. (1966). 'The Life History of *Odostomia bisuturalis,* and *Odostomia* Spermatophores (Gastropoda, Pyramidellidae)', *Year Book of the American Philosophical Society, (1966),* pp. 368-70
Robson, E.M. and Williams, I.C. (1971). 'Relationships of Some Species of Digenea with the Marine Prosobranch *Littorina littorea* (L.) 2. The Effect of Larval Digenea on the Reproductive Biology of *L. littorea*', *Journal of Helminthology, 45,* pp. 145-59
Rose, R.A. and Hoegh-Guldberg, I.O. (1982). 'A Brood-protecting Nudibranch with Pelagic Lecithotrophic Development', *Journal of Molluscan Studies, 48,* pp. 231-2
Rudman, W.B. (1971). 'Structure and Functioning of the Gut in the Bullomorpha (Opisthobranchia). Part I. Herbivores', *Journal of Natural History, 5,* pp. 647-75
Rudman, W.B. (1972a). 'Structure and Functioning of the Gut in the Bullomorpha (Opisthobranchia). Part 3. Philinidae', *Journal of Natural History, 6,* pp. 459-74
Rudman, W.B. (1972b). 'Structure and Functioning of the Gut in the Bullomorpha (Opisthobranchia). Part 4. Aglajidae', *Journal of Natural History, 6,* pp. 547-60
Runham, W.W. In press. 'Mollusca', in Adiyodi, K.G. and Adiyodi, R.G., (eds), *Reproductive Biology of Invertebrates, Vols 3, 5, 6*
Runham, W.W. and Isarankura, K. (1966). 'Studies on Radula Replacement', *Malacologia, 5,* p. 73
Runham, W.W. and Thornton, P.R. (1967). 'Mechanical Wear of the Gastropod Radula: a Scanning Electron Microscope Study', *Journal of Zoology, London, 153,* pp. 445-52
Runham, W.W., Thornton, P.R., Shaw, D.A. and Wayte, R.C. (1969). 'The Mineralization and Hardness of the Radular Teeth of the Limpet *Patella vulgata* L.', *Zeitschrift für Zellforschung und mikroskopische Anatomie, 99,* pp. 608-26
Sandeen, M.I., Stephens, G.C. and Brown, F.A. (1954). 'Persistent Daily and Tidal Rhythms of Oxygen Consumption in Two Species of Marine Snails', *Physiological Zoology, 27,* pp. 350-6
Sandison, E.E. (1966). 'The Oxygen Consumption of Some Intertidal Gastropods in Relation to Zonation', *Journal of Zoology, London, 149,* pp. 163-73
Scheller, R.H. and McAllister, L.B. (1983). 'Molecular Cloning of a Multigene Family Encoding Neuropeptides which Govern Egg-laying in *Aplysia*', in Lever, J. and Boer, M.H. (eds), *Molluscan Neuro-endocrinology.* Proceedings

of the International Minisymposium on Molluscan Endocrinology, held in the Department of Biology, Free University, Amsterdam, Netherlands, 16-20 August 1982. North-Holland Publishing Company, Amsterdam, pp. 38-44

Scheltema, R.S. (1968). 'Dispersal of Larvae by Equatorial Ocean Currents and its Importance to the Zoogeography of Shoal-water Tropical Species', *Nature (London), 217*, pp. 1159-62

Scheltema, R.S. (1978). 'On the Relationship between Dispersal of Pelagic Veliger Larvae and the Evolution of Marine Prosobranch Gastropods', in Battaglia B. and Beardmore, J.A. (eds), *Marine Organisms: Genetics, Ecology and Evolution*, NATO Conference Series. Series IV: Marine Sciences, Plenum Press, New York, pp. 303-22

Seed, R. (1978). 'Observations on the Adaptive Significance of Shell Shape and Body Form in Dogwhelks (*Nucella lapillus* (L.)) from N. Wales', *Nature in Wales, 16*, pp. 111-22

Seifert, von R. (1935). 'Bemerkungen zur Artunterscheidung der deutschen Brackwasser-Hydrobien', *Zoologischer Anzeiger, 110*, pp. 233-9

Seilacher, A. von (1959). 'Schnecken im Brandungssand', *Natur und Volk, 89*, pp. 359-66

Shackleton, N.J. (1973). 'Oxygen Isotope Analysis as a Means of Determining Season of Occupation of Prehistoric Midden Sites', *Archaeometry, 15(1)*, pp. 133-41

Shimek, R.L. (1981). '*Neptunea pribiloffensis* (Dall, 1919) and *Tealia crassicornis* (Muller, 1776): on a Snail's Use of Babysitters', *Veliger, 24*, pp. 62-6

Shimek, R.L. (1984). 'The Diets of Alaskan *Neptunea*', *Veliger, 26*, pp. 274-81

Shirley, T.C. and Findley, A.M. (1978). 'Circadian Rhythm of Oxygen Consumption in the Marsh Periwinkle *Littorina irrorata* (Say, 1822)', *Comparative Biochemistry and Physiology, 59A*, pp. 339-42

Shumway, S.E. (1978). 'The Effects of Fluctuating Salinity on Respiration in Gastropod Molluscs', *Comparative Biochemistry and Physiology, 63A*, pp. 279-83

Sibly, R. and Calow, P. (1982). 'Asexual Reproduction in Protozoa and Invertebrates', *Journal of Theoretical Biology, 96*, pp. 401-24

Sminia, T. (1972). 'Structure and Function of Blood and Connective Tissue Cells of the Fresh-water Pulmonate *Lymnaea stagnalis* Studied by Electron Microscopy and Enzyme Histochemistry', *Zeitschrift für Zellforschung und mikroskopische Anatomie, 130*, pp. 497-526

Smith, B.S. (1981). 'Male Characteristics in Female *Nassarius obsoletus*: Variations Related to Locality, Season and Year', *Veliger, 23*, pp. 212-16

Smith, D.A. and Sebens, K.P. (1983). 'The Physiological Ecology of Growth and Reproduction in *Onchidoris aspersa* (Alder and Hancock) (Gastropoda: Nudibranchia)', *Journal of Experimental Marine Biology and Ecology, 72*, pp. 287-304

Smith, T.B. (1984). 'Ultrastructure and Function of the Proboscis in *Melanella alba* (Gastropoda: Eulimidae)', *Journal of the Marine Biological Association of the United Kingdom, 64*, pp. 503-12

Spight, T.M. (1975). 'Factors Extending Gastropod Embryonic Development and their Selective Cost', *Oecologia (Berlin), 21*, pp. 1-16

Spight, T.M. (1976). 'Hatching Size and the Distribution of Nurse Eggs among Prosobranch Embryos', *Biological Bulletin, 150*, pp. 491-9

Spight, T.M. and Emlen, J. (1976). 'Clutch Sizes of Two Marine Snails with a Changing Food Supply', *Ecology, 57*, pp. 1162-78

Stearns, S.C. (1976). 'Life-history Tactics: a Review of the Ideas', *Quarterly Review of Biology, 51*, pp. 3-47

Steneck, R.S. (1982). 'A Limpet-Coralline Alga Association: Adaptations and Defences between a Selective Herbivore and its Prey', *Ecology, 63*, pp. 507-22

Steneck, R.S. and Watling, L. (1982). 'Feeding Capabilities and Limitation of Hebivorous Molluscs: a Functional Group Approach', *Marine Biology, 68*, pp. 299-319

Stephenson, T.A. and Stephenson, A. (1972). *Life between Tide Marks on Rocky Shores*, W.H. Freeman, San Francisco

Stickle, W.B. (1971). 'The Metabolic Effects of Starving *Thais lamellosa* Immediately after Spawning', *Comparative Biochemistry and Physiology, 40A*, pp. 627-34

Stickle, W.B. (1973). 'The Reproductive Physiology of the Intertidal Prosobranch *Thais lamellosa* (Gmelin). I. Seasonal Changes in the Rate of Oxygen Consumption and Body Component Indices', *Biological Bulletin, 144*, pp. 511-24

Stickle, W.B. (1975). 'The Reproductive Physiology of the Intertidal Prosobranch *Thais lamellosa* (Gmelin). 2. Seasonal Changes in Biochemical Composition', *Biological Bulletin of the Marine Biological Laboratory, Woods Hole, 148*, pp. 448-60

Stickle, W.B. and Bayne, B.L. (1982). 'Effects of Temperature and Salinity on Oxygen Consumption and Nitrogen Excretion in *Thais (Nucella) lapillus* (L.)', *Journal of Experimental Marine Biology and Ecology, 58*, pp. 1-17

Stimson, J. (1970). 'Territorial Behavior of the Owl Limpet, *Lottia gigantea*', *Ecology, 51*, pp. 113-18

Stirts, H.M. and Clark, K.B. (1980). 'Effects of Temperature on Products of Symbiotic Chloroplasts in *Elysia tuca* Marcus (Opisthobranchia: Ascoglossa)', *Journal of Experimental Marine Biology and Ecology, 43*, pp. 39-47

Strathmann, R.R. and Chaffee, C. (1984). 'Constraints on Egg Masses. II. Effect of Spacing, Size, and Number of Eggs on Ventilation of Masses of Embryos in Jelly, Adherent Groups, or in Thin-walled capsules', *Journal of Experimental Marine Biology and Ecology, 84*, pp. 85-93

Strathmann, R.R. and Strathmann, M.F. (1982). 'The Relationship between Adult Size and Brooding in Marine Invertebrates', *American Naturalist, 119*, pp. 91-101

Sutherland, J. (1970). 'Dynamics of High and Low Populations of the Limpet, *Acmaea scabra* (Gould)', *Ecological Monographs, 40*, pp. 169-88

Taylor, J.D. (1976). 'Habitats, Abundance and Diets of Muricacean Gastropods at Aldabra Atoll', *Zoological Journal of the Linnean Society, 59*, pp. 155-93

Taylor, J.D. (1984). 'A Partial Food Web Involving Predatory Gastropods on a Pacific Fringing Reef', *Journal of Experimental Marine Biology and Ecology, 74*, pp. 273-90

Taylor, J.D. and Layman, M. (1972). 'The Mechanical Properties of Bivalve (Mollusca) Shell Structures', *Paleontology, 15*, 73-87

Taylor, J.D. and Taylor, C.N. (1977). 'Latitudinal Distribution of Predatory Gastropods on the Eastern Atlantic Shelf', *Journal of Biogeography, 4*, pp. 73-81

Taylor, J.D., Morris, N.J. and Taylor, C.N. (1980). 'Food Specialization and the Evolution of Predatory Prosobranch Gastropods', *Paleontology, 23*, pp. 375-409

Thompson, T.E. (1958). 'The Influence of Temperature on Spawning in *Alderia proxima* (A.H.) (Gastropoda: Nudibranchia)', *Oikos, 9*, pp. 246-52

Thompson, T.E. (1976). *Biology of Opisthobranch Molluscs, Vol. 1*, Ray Society, London

Thompson, T.E. and Brown, G.H. (1984). *Biology of Opisthobranch Molluscs, Vol. II*, Ray Society, London

Thorson, G. (1946). 'Reproduction and Larval Development of Danish Marine Bottom Invertebrates, with Special Reference to the Planktonic Larvae in the Sound (Oresund)', *Meddelelser fra Danmarks Fisken og Havundersogelser Ser. Plankton, 4*, pp. 1-523

Thorson, G. (1950). 'Reproductive and Larval Ecology of Marine Bottom Invertebrates', *Biological Reviews, 25*, pp. 1-45

Thorson, G. (1966). 'Some Factors Influencing the Recruitment and Establishment of Marine Benthic Communities', *Netherlands Journal of Sea Research, 3*, pp. 267-93

Todd, C.D. (1981). 'The Ecology of Nudibranch Molluscs', *Oceanography and Marine Biology Annual Review, 19*, pp. 141-234

Todd, C.D. and Doyle, R.W. (1981). 'Reproductive Strategies of Marine Benthic Invertebrates: a Settlement-timing Hypothesis', *Marine Ecology Progress Series, 4*, pp. 75-83

Townsend, C.R. and Calow, P. (eds) (1981). *Physiological Ecology: an Evolutionary Approach to Resource Use*, Blackwell Scientific Publications, Oxford

Townsend, C.R. and Hughes, R.N. (1981). 'Maximizing Net Energy Returns from Foraging', in Townsend, C.R. and Calow, P. (eds), *Physiological Ecology: an Evolutionary Approach to Resource Use*, Blackwell Scientific Publications, Oxford, pp. 86-108

Uma Devi, V., Prabhakara Rao, Y. and Prasada Rao, D.G.V. (1984). 'Anaerobic Response of a Tropical Intertidal Gastropod *Morula granulata* (Duclos) to Low Salinities and Fresh Water', *Journal of Experimental Marine Biology and Ecology, 84*, pp. 179-89

Underwood, A.J. (1974). 'The Reproductive Cycles and Geographical Distribution of Some Common Eastern Australian Prosobranchs (Mollusca: Gastropoda)', *Australian Journal of Marine and Freshwater Research, 25*, pp. 63-88

Underwood, A.J. (1976). 'Food Competition between Age-classes in the Intertidal Neritacean *Nerita atramentosa* Reeve (Gastropoda: Prosobranchia)', *Journal of Experimental Marine Biology and Ecology, 23*, pp. 145-54

Underwood, A.J. (1979). 'The Ecology of Intertidal Gastropods', *Advances in Marine Biology, 16*, pp. 111-210

Underwood, A.J. (1980). 'The Effects of Grazing by Gastropods and Physical Factors on the Upper Limits of Distribution of Intertidal Macroalgae', *Oecologia (Berlin), 46*, pp. 201-13

Underwood, A.J. (1984). 'Microalgal Food and the Growth of the Intertidal Gastropods *Nerita atramentosa* Reeve and *Bembicium nanum* (Lamarck) at Four Heights on a Shore', *Journal of Experimental Marine Biology and Ecology, 79*, pp. 277-91

Underwood, A.J., Denley, E.J. and Moran, M.J. (1983). 'Experimental Analysis of the Structure and Dynamics of Mid-shore Rocky Intertidal Communities in New South Wales', *Oecologia (Berlin), 56*, pp. 202-19

Vahl, O. (1983). 'Mucus Drifting in the Limpet *Helcion (= Patina) pellucida* (Prosobranchia, Patellidae)', *Sarsia, 68*, pp. 209-11

Valentine, J.W. (1973). *Evolutionary Paleoecology of the Marine Biosphere*, Prentice-Hall, Englewood Cliffs, New Jersey

Vance, R.R. (1973). 'On Reproductive Strategies in Marine Benthic Invertebrates', *American Naturalist, 107*, pp. 339-52

Verderber, G.W., Cook, S.B. and Cook, C.B. (1983). 'The Role of the Homing Scar in Reducing Water Loss during Aerial Exposure of the Pulmonate Limpet *Siphonaria alternata* (Say)', *Veliger, 25*, pp. 235-43

Vermeij, G.J. (1966). 'Interoceanic Differences in Vulnerability of Shelled Prey to Crab Predation', *Nature (London), 260*, pp. 135-6

Vermeij, G.J. (1978). *Biogeography and Adaptation: Patterns in Marine Life*, Harvard University Press, Cambridge, Mass.

Vermeij, G.J. (1982). 'Phenotypic Evolution in a Poorly Dispersing Snail after Arrival of a Predator', *Nature (London), 299*, pp. 349-50

Walsby, J.R., Morton, J.E. and Croxall, J.P. (1973). 'The Feeling Mechanism and

Ecology of the New Zealand Pulmonate Limpet, *Gadinalea nivea*', *Journal of Zoology, 171*, pp. 257-83
Warner, R.R. (1975). 'The Adaptive Significance of Sequential Hermaphroditism in Animals', *American Naturalist, 109*, pp. 61-82
Warwick, T. (1982). 'A Method of Maintaining and Breeding Members of the *Littorina saxatilis* (Olivi) Species Complex', *Journal of Molluscan Studies, 48*, pp. 368-70
Watanabe, J.M. (1984). 'Food Preferences, Food Quality and Diets of Three Herbivorous Gastropods (Trochidae = *Tegula*) in a Temperate Kelp Forest Habitat', *Oecologia* (Berlin), *62*, pp. 47-52.
Watson, D.C. and Norton, T.A. (1985). 'Dietary Preferences of the Common Periwinkle *Littorina littorea*', *Journal of Experimental Marine Biology and Ecology*, (in press)
Wells, R.A. (1980). 'Activity Pattern as a Mechanism of Predator Avoidance in Two Species of Acmaeid Limpet', *Journal of Experimental Marine Biology and Ecology, 48*, pp. 151-68
West, H.H., Harrigan, J.F. and Pierce, S.K. (1984). 'Hybridization of Two Populations of a Marine Opisthobranch with Different Developmental Patterns', *Veliger, 26*, pp. 199-206
Wigham, G.D. (1976). 'Feeding and Digestion in the Marine Prosobranch *Rissoa parva* (da Costa)', *Journal of Molluscan Studies, 42*, pp. 74-94
Wilbur, K.M. (ed. in chief) (1983). *The Mollusca, Vols. 1-5*, Academic Press, London
Wilbur, K.M. (1984). *The Mollusca, Vol. 6*, Academic Press, London
Wilkins, N.P. and O'Regan, D. (1980). 'Genetic Variation in Sympatric Sibling Species of *Littorina*', *Veliger, 22*, pp. 355-9
Williams, E.E. (1970). 'Seasonal Variations in the Biochemical Composition of the Edible Periwinkle *Littorina littorea* (L.)', *Comparative Biochemistry and Physiology, 33*, pp. 655-61
Williams, J.C. and Ellis, C. (1975). 'Movements of the Common Periwinkle, *Littorina littorea* (L.), on the Yorkshire Coast in Winter and the Influence of Infection with Larval Digenea', *Journal of Experimental Marine Biology and Ecology, 17*, pp. 47-58
Williamson, P. and Kendall, M.A. (1981). 'Population Age Structure and Growth of the Trochid *Monodonta lineata* Determined from Shell Rings', *Journal of the Marine Biological Association of the United Kingdom, 61*, pp. 1011-26
Wolcott, T.G. (1973). 'Physiological Ecology and Intertidal Zonation in Limpets (Acmaea): a Critical Look at "Limiting factors"', *Biological Bulletin, 145*, pp. 389-422
Wood, L. (1968). 'Physiological and Ecological Aspects of Prey Selection by the Marine Gastropod *Urosalpinx cinerea* (Prosobranchia, Muricidae)', *Malacologia, 6*, pp. 267-320
Wright, J.R. and Hartnoll, R.G. (1981). 'An Energy Budget for a Population of the Limpet *Patella vulgata*', *Journal of the Marine Biological Association of the United Kingdom, 61*, pp. 627-46
Wright, W.G. and Lindberg, D.R. (1979). 'A Non-fatal Method of Sex Determination for Patellacean Gastropods', *Journal of the Marine Biological Association of the United Kingdom, 59*, p. 803
Yamaguchi, M. (1977). 'Shell Growth and Mortality Rates in the Coral Reef Gastropod *Cerithium nodulosum* in Pago Bay, Guam, Mariana Islands', *Marine Biology, 44*, pp. 249-63
Yonge, C.M. (1926). 'Ciliary Feeding Mechanisms in the Thecosomatous Pteropods', *Journal of the Linnean Society, 36*, pp. 417-29
Yonge, C.M. (1930). 'The Crystalline Style of the Mollusca and a Carnivorous Habit Cannot Normally Coexist', *Nature (London), 125*, pp. 444-5

Yonge, C.M. and Thompson, T.E. (1976). *Living Marine Molluscs*, Collins, London
Zeldis, J.R. and Boyden, C.R. (1979). 'Feeding Adaptations of *Melagraphia aethiops* (Gmelin), an Intertidal Trochid Mollusc', *Journal of Experimental Marine Biology and Ecology*, *40*, pp. 267-83
Zinsmeister, W.J. and Emerson, W.K. (1979) 'The Role of Passive Dispersal in the Distribution of Hemipelagic Invertebrates, with Examples from the Tropical Pacific Ocean', *Veliger*, *22*, pp. 32-40
Zipser, E. and Vermeij, G.J. (1978). 'Crushing Behavior of Tropical and Temperate Crabs', *Journal of Experimental Marine Biology and Ecology*, *31*, pp. 155-72

INDEX